Thomas Bingley

Stories About the Instinct of Animals

their characters and habits

Thomas Bingley

Stories About the Instinct of Animals

their characters and habits

ISBN/EAN: 9783337238469

Printed in Europe, USA, Canada, Australia, Japan

Cover: Foto ©berggeist007 / pixelio.de

More available books at **www.hansebooks.com**

STORIES

ABOUT THE

INSTINCT OF ANIMALS,

THEIR CHARACTERS, AND HABITS.

BY THOMAS BINGLEY.

EMBELLISHED WITH ENGRAVINGS, FROM DRAWINGS BY T. LANDSEER.

BOSTON:
CROSBY, NICHOLS, LEE AND COMPANY.
1861.

CONTENTS.

CHAPTER I.

Uncle Thomas resumes his Stories about the Instinct of Animals.—Tells about the Horse, and of the Immense Herds which are to be found on the Plains of South America; of their Capture by means of the Lasso; the Arab and his Mare; the Gadshill Robber; the Benevolent Planter; the Lawyer-Highwayman; as well as several other Curious Stories about the Intelligence, Affection, and Docility of the Horse..Page 9

CHAPTER II.

Uncle Thomas tells about the Beaver, and the Singular Manner in which it constructs a Dam to confine the Waters of the River; and about the Hut which it builds for its Habitation. He tells also about the Curious Nests of the Sociable Grosbeak; and gives a Long and Entertaining Account of the White Ant of Africa; its Extraordinary Nest; and the Important Part which it acts in the Economy of Nature.............................29

CHAPTER III.

Uncle Thomas describes the Manner in which W... Elephants are caught, and relates some Curious Stories of ... Affection, and Intelligence of the Elephant.........

CONTENTS.

CHAPTER IV.

Uncle Thomas introduces to the Notice of the Young Folks the Ettrick Shepherd's Stories about Sheep ; and tells them some Interesting Stories about the Goat, and its Peculiarities.......71

CHAPTER V.

Uncle Thomas relates some Very Remarkable Stories about the Cat ; points out to the Boys the Connexion subsisting between the Domestic Cat and the Lion, Tiger, &c., and tells them some Stories about the Gentleness, as well as the Ferocity of these Animals...89

CHAPTER VI.

Uncle Thomas tells about the Tiger ; its Ferocity and Power ; and of the Curious Modes which are adopted for its Capture and Destruction.—Also about the Puma or American Lion, and introduces some Hunting Scenes in North and South America, with other Interesting and Entertaining Adventures...123

CHAPTER VII.

Uncle Thomas tells about the Migrating Instinct of Animals.- Of the House Swallow of England ; and the Esculent Swallow, whose Nest is eaten by the Chinese.—He tells also about the Passenger Pigeon of America ; of the Myriads which are found in various parts of the United States; of the Land-Crab and its Migrations, and of those of the Salmon and the Common Herring................................144

CONTENTS. vii

CHAPTER VIII.

Uncle Thomas tells about the Baboons, and their Plundering Excursions to the Gardens at the of Good Hope, Calsoaep about Le Vaillant's Baboon, Kees, and his Peculiarities; the American Monkeys; and relates an Amusing Story about a Young Monkey deprived of its Mother, putting itself under the Fostering Care of a Wig-Block..................................174

CHAPTER IX.

Uncle Thomas concludes Stories about Instinct with several Interesting Illustrations of the Affections of Animals, particularly of the Instinct of Maternal Affection, in the course of which he narrates the Story of the Cat and the Black-Bird; the Squirrel's Nest; the Equestrian Friends; and points out the Beneficent Care of Providence in implanting in the Breasts of each of his Creatures the Instinct which is necessary for its Security and Protection............................193

STORIES

ILLUSTRATIVE OF THE

INSTINCT OF ANIMALS.

CHAPTER I.

Uncle Thomas resumes his Stories about the Instinct of Animals.—Tells about the Horse, and of the Immense Herds which are to be found on the Plains of South America; of their Capture by means of the Lasso; the Arab and his Mare; the Gadshill Robber; the Benevolent Planter; the Lawyer-Highwayman; as well as several other Curious Stories about the Intelligence, Affection, and Docility of the Horse.

"COME away, boys, I am glad to see you again! Since I last saw you I have made an extensive tour, and visited some of the most romantic and picturesque scenery in England. One day I may give you an account of what I saw, and describe to you the scenes which I visited; but I must deny myself this pleasure at present. I promised, at our next meeting, to tell you some Tales about the Instinct of Animals; and I propose to begin with the Horse. I like to interest you with those animals with which you are familiar, and to draw out your sympathies towards them. After the Stories about Dogs which I told you, some of them exhibiting that fine animal in such an amiable and affectionate character, I am sure it must assume a new interest in your mind. Such instances of fidelity and attachment could not fail to impress you with a higher opinion of the animal than you before possessed, and show that kindness and good treatment even to a brute are not without their reward.

"I wish to excite the same interest towards the other animals which, I hope, I have effected towards the Dog. Each, you will find, has been endowed by its Creator with particular instincts, to fit it for the station which it was intended to occupy in the great system of Nature. Some of them are wild and ferocious, while others are quiet and inoffensive; the former naturally repel us, while those of the latter class as naturally attract our regard, although, properly speaking, each ought equally to interest us, in as far as it fulfils the object of its being.

"But I know you like stories better than lectures, so I will not tire you by lecturing, but will at once proceed to tell some stories about Horses, which I have gathered for you."

"Oh no, Uncle Thomas, we never feel tired of listening to you; we know you have always something curious to tell us."

"Well, then, Frank, to begin at once with THE HORSE.

"In several parts of the world there are to be found large herds of wild horses. In South America, in particular, the immense plains are inhabited by them, and, it is said, that so many as ten thousand are sometimes found in a single herd. These flocks are always preceded by a leader, who directs their motions; and such is the regularity with which they perform their movements, that it seems almost as if they could not be surpassed by the best trained cavalry.

"It is extremely dangerous for travellers to encounter a herd of this description. When they are unaccustomed to the sight of such a mass of creatures, they cannot help feeling greatly alarmed at their rapid and apparently irresistible approach. The trampling of the animals sounds like the loudest thunder; and such is the rapidity and impetuosity of their advance, that it seems to threaten instant destruction. Suddenly, however, they sometimes stop short, utter a loud and piercing neighing, and, with a rapid wheel in an opposite course, altogether disappear. On such occasions, however, it requires all the care of the traveller to prevent his horses from breaking loose, and escaping with the wild herd.

"In those countries where horses are so plentiful, the inhabitants do not take the trouble to rear them, but, whenever they want one, mount upon an animal which has been accustomed to the sport, and gallop over the plain towards the herd, which is readily found at no great distance. Gradually he approaches some stragglers from the main body, and, having selected the horse which he wishes to possess, he dexterously throws the *lasso* (which is a long rope with a running noose, and which is firmly fixed to his saddle,) in such a manner as to entangle the animal's hind legs; and, with a sudden turn of his horse, he pulls it over on its side. In an instant he jumps off his horse, wraps his *poncho*, or cloak, round the captive's head, forces a bit into its mouth, and straps a saddle upon its back. He then removes the poncho, and the animal starts on its feet. With equal quickness the hunter leaps into the saddle; and, in spite of the contortions and kickings of his captive, keeps his seat, till, having wearied itself out with its vain efforts, it submits to the discipline of its captor, who seldom fails to reduce it to complete obedience."

"That is very dexterous indeed, Uncle Thomas; but surely all horses are not originally found in this wild state. I have heard that the Arabians are famous for rearing horses."

"Arabia has, for a long time, been the country noted for the symmetry and speed of its horses: so much attention has been paid to the breeding of horses in our own country, however, for the race-course as well as the hunting-field, that the English horses are now almost unequalled, both for speed and endurance.

"It is little wonder, however, that the Arabian horse should be the most excellent, considering the care and attention which it receives, and the kindness and consideration with which it is treated. One of the best stories which I ever heard of the love of an Arabian for his steed, is that related of an Arab from whom one of our envoys wished to purchase his horse.

"The animal was a bright bay mare, of extraordinary shape and beauty; and the owner, proud of its appearance and qualities, paraded it before the envoy's tent until it attracted his attention. On being asked if he would sell her, 'What will you give me?' was the reply. 'That depends upon her age; I suppose she is past five?' 'Guess again,' said he. 'Four?' 'Look at her mouth,' said the Arab, with a smile. On examination she was found to be rising three. This, from her size and symmetry, greatly enhanced her value. The envoy said, 'I will give you fifty tomans' (a coin nearly of the value of a pound sterling). 'A little more, if you please,' said the fellow, somewhat entertained. 'Eighty—a hundred.' He shook his head and smiled. The officer at last came to two hundred tomans. 'Well,' said the Arab, 'you need not tempt me farther. You are a rich elchee (nobleman); you have fine horses, camels, and mules, and I am told you have loads of silver and gold. Now,' added he, 'you want my mare, but you shall not have her for all you have got.' He put spurs to his horse, and was soon out of the reach of temptation.

"Swift as the Arabian horses are, however, they are frequently matched by those of our own country. I say nothing about *race horses*, because, though some of them are recorded to have run at an amazing speed, the effort is generally continued for but a short time. Here is an instance of speed in a horse which saved its unworthy master from the punishment due to his crime.

"One morning about four o'clock a gentleman was stopped, and robbed by a highwayman named Nicks, at Gadshill, on the west side of Chatham. He was mounted on a bay mare of great speed and endurance, and as soon as he had accomplished his purpose, he instantly started for Gravesend, where he was detained nearly an hour by the

THE GADSHILL ROBBER. 17

difficulty of getting a boat. He employed the interval to advantage however in baiting his horse. From thence he got to Essex and Chelmsford, where he again stopped about half an hour to refresh his horse. He then went to Braintree, Bocking, Weathersfield, and over the Downs to Cambridge, and still pursuing the cross roads, he went by Fenney and Stratford to Huntingdon, where he again rested about half an hour. Proceeding now on the north road, and at a full gallop most of the way, he arrived at York the same afternoon, put off his boots and riding clothes, and went dressed to the bowling-green, where, among other promenaders, happened to be the Lord Mayor of the city. He there studied to do something particular, that his lordship might remember him, and asking what o'clock it was, the mayor informed him that it was a quarter past eight. Notwithstan ding all these precautions, however, he was discovered, and tried for the robbery; he rested his defence on the fact of his being at York at such a time. The gentleman swore positively to the time and place at which the robbery was committed, but on the other hand, the proof was equally clear that the prisoner was at York at the time specified. The jury acquitted him on the supposed impossibility of his having got so great a distance from Kent by the time he was seen in the bowling-green. Yet he was the highwayman."

"So that he owed his safety to the speed of his horse, Uncle Thomas."

"He did so, Harry. The horse can on occasion swim about as well as most animals, yet it never takes to the water unless urged to do so. There is a story about a horse saving the lives of many persons who had suffered shipwreck by being driven upon the rocks at the Cape of Good Hope, which, I am sure, will interest you as much for the perseverance and docility of the animal, as for the benevolence and intrepidity of its owner.

"A violent gale of wind setting in from north and north-west, a vessel in the roads dragged her anchors, was forced on the rocks, and bilged; and while the greater part of the crew fell an immediate sacrifice to the waves, the remainder were seen from the shore struggling for their lives, by clinging to the different pieces of the wreck. The sea ran dreadfully high, and broke over the sailors with such amazing fury, that no boat whatever could venture off to their assistance. Meanwhile a planter, considerably advanced in life, had come from his farm to be a spectator of the shipwreck; his heart was melted at the sight of the unhappy seamen, and knowing the bold and enterprizing spirit of his horse, and his particular excellence as a swimmer, he instantly determined to make a desperate effort for their deliverance. He alighted, and blew a little brandy into his horse's nostrils, and again seating himself in the saddle, he instantly pushed into the midst of the breakers. At first both disappeared, but it was not long before they floated on the surface, and swam up to the wreck; when taking with him two men, each of whom held by one of his boots, he brought them safe to shore. This perilous expedition he repeated no seldomer than seven times, and saved fourteen lives; but on his return the eighth time, his horse being much fatigued, and meeting a most formidable wave, he lost his balance, and was overwhelmed in a moment. The horse swam safely to land, but his gallant rider sank to rise no more."

"That was very unfortunate, Uncle Thomas. I suppose the planter had been so fatigued with his previous exertions, that he had not strength to struggle with the strong waves."

"Very likely, indeed, Harry. I dare say the poor animal felt the loss of his kind owner very much, for the horse soon becomes attached to his master, and exhibits traits of intelligence and fidelity, certainly, not surpassed by those of any other animal: for instance,—A gentleman, who was one dark night riding home through a wood, had the misfortune to strike his head against the branch of a tree, and fell from his horse stunned by the blow. The noble animal immediately returned to the house they had left, which stood about a mile distant. He found the door closed,—the family had retired to bed. He pawed at it, however, till one of them, hearing the noise, arose and opened it, and, to his surprise, saw the horse of his friend. No sooner was the door opened than the horse turned round as if it wished to be followed; and the man, suspecting there was something wrong, followed the animal, which led him directly to the spot where its wounded master lay on the ground.

"There is another story of a somewhat similar description in which a horse saved his master from perishing among the snow. It happened in the North of Scotland.

"A gentleman connected with the Excise was returning home from one of his professional journies. His way lay across a range of hills, the road over which was so blocked up with snow as to leave all trace of it indiscernible. Uncertain how to proceed, he resolved to trust to his horse, and throwing loose the reins, allowed him to choose his course. The animal proceeded cautiously, and safely for some time, till coming to a ravine, horse and rider sunk in a snow-wreath several fathoms deep.

"Stunned by the suddenness and depth of the descent, the gentleman lay for some time insensible. On recovering, he found himself nearly three yards from the dangerous spot, with his faithful horse standing over him and licking the snow from his face. He accounts for his extrication, by supposing that the bridle must have been attached to his person, but so completely had he lost all sense of consciousness, that beyond the bare fact as stated, he had no knowledge of the means by which he made so remarkable an escape."

"It was at any rate very kind in the horse to clear away the snow, Uncle Thomas."

"No doubt of it, John, and perhaps he owed his life quite as much to this act of kindness as to being pulled out of the ravine. He might have been as certainly choked by the snow out of it as in it. Sometimes the horse becomes much attached to the animals with which it associates, and its feelings of friendship are as powerful as those of the dog. A gentleman of Bristol had a greyhound which slept in the same stable, and contracted a very great intimacy with a fine hunter. When the dog was taken out, the horse neighed wistfully after him, and seemed to long for its return; he welcomed him home with a neigh; the greyhound ran up to the horse and licked him; the horse, in return, scratched the greyhound's back with his teeth. On one occasion, when the groom had the pair out for exercise, a large dog attacked the greyhound, bore him to the ground, and seemed likely to worry him, when the horse threw back his ears, rushed forward, seized the strange dog by the back, and flung him to a distance which so terrified the aggressor, that he at once desisted and made off."

" That was very kind, Uncle Thomas. I like to hear of such instances of friendship between animals."

" Such a docile animal as the horse can readily be trained to particular habits, and does not readily forget them, however disreputable. There is an odd story to illustrate this.

" About the middle of last century, a Scottish lawyer had occasion to visit the metropolis. At that period such journies were usually performed on horseback, and the traveller might either ride post, or, if willing to travel economically, he bought a horse, and sold him at the end of his journey. The lawyer had chosen the latter mode of travelling, and sold the animal on which he rode from Scotland as soon as he arrived in London. With a view to his return, he went to Smithfield to purchase a horse. About dusk a handsome one was offered, at so cheap a rate that he suspected the soundness of the animal, but being able to discover no blemish, he became the purchaser.

"Next morning, he set out on his journey, the horse had excellent paces, and our traveller, while riding over the few first miles, where the road was well frequented, did not fail to congratulate himself on his good fortune, which had led him to make so advantageous a bargain.

"They arrived at last at Finchley Common, and at a place where the road ran down a slight eminence, and up another, the lawyer met a clergyman driving a one-horse chaise. There was nobody within sight, and the horse by his conduct instantly discovered the profession of his former owner. Instead of pursuing his journey, he ran close up to the chaise and stopt it, having no doubt but his rider would embrace so fair an opportunity of exercising his calling. The clergyman seemed of the same opinion, produced his purse unasked, and assured the astonished lawyer that it was quite unnecessary to draw his pistol, as he did not intend to offer any resistance. The traveller rallied his horse, and with many apologies to the gentleman he had so innocently and unwillingly affrighted, pursued his journey.

"They had not proceeded far when the horse again made the same suspicious approach to a coach, from the window of which a blunderbuss was levelled, with denunciations of death and destruction to the hapless and perplexed rider. In short, after his life had been once or twice endangered by the suspicions to which the conduct of his horse gave rise, and his liberty as often threatened by the peace-officers, who were disposed to apprehend him as a notorious highwayman, the former owner of the horse, he was obliged to part with the inauspicious animal for a trifle, and to purchase one less beautiful, but not accustomed to such dangerous habits."

"Capital, Uncle Thomas! I should have liked to have seen the perplexed look of the poor lawyer, when he saw the blunderbuss make its appearance at the carriage window!"

"There is one other story about the horse, showing his love for his master, and the gentleness of his character. A horse which was re markable for its antipathy to strangers, one evening, while bearing his master home from a jovial meeting, became disburthened of his rider, who, having indulged rather freely, soon went to sleep on the ground. The horse, however, did not scamper off, but kept faithful watch by his prostrate master till the morning, when the two were perceived about sunrise by some labourers. They approached the gentleman, with the intention of replacing him on his saddle, but every attempt on their part was resolutely opposed by the grinning teeth and ready heels of the horse, which would neither allow them to touch his master, nor suffer himself to be seized till the gentleman himself awoke from his sleep. The same horse, among other bad propensities, constantly resented the attempts of the groom to trim its fetlocks. This circumstance happened to be mentioned by its owner in conversation, in the presence of his youngest child, a very few years old, when he defied any man to perform the operation singly. The father next day, in passing through the stable-yard, beheld with the utmost distress, the infant employed with a pair of scissors in clipping the fetlocks of the hind-legs of this vicious hunter—an operation which had been always hitherto performed with great danger even by a number of men. Instead, however, of exhibiting his usual vicious disposition, the horse, in the present case, was looking with the greatest complacency on the little groom, who soon after, to the very great relief of his father, walked off unhurt."

CHAPTER II.

Uncle Thomas tells about the Beaver, and the Singular Manner in which it Constructs a Dam to confine the Waters of the River; and about the Hut which it builds for its Habitation. He tells also about the Curious Nests of the Sociable Grosbeak; and gives a Long and Entertaining Account of the White Ant of Africa; its Extraordinary Nest; and the Important Part which it acts in the Economy of Nature.

"Good evening, Boys! I am going to tell you about a very singular animal to-night—singular both in its conformation and its habits. I allude to the Beaver."

"Oh, we shall be so glad to hear about the Beaver, Uncle Thomas. I have sometimes wondered what sort of an animal it is. It is of its skin that hats are made—is it not?"

"It is so, Harry—at least it is of the fur with which its skin is covered. I must tell you about the manufacture of hats at some other time. Our business at present is with the Beaver itself. I think we shall get on better by confining our attention to the animal now, and examine into its habits and instincts."

"Very well, Uncle Thomas, we are all attention."

"The Beaver, which is now only to be found in the more inaccessible parts of America, and the more northern countries of Europe, affords a curious instance of what may be called a compound structure. It has the fore-feet of a land animal, and the hind ones of an aquatic one— the latter only being webbed. Its tail is covered with scales like a fish, and serves to direct its course in the water, in which it spends much of its time.

"On the rivers where they abound, they form societies sometimes consisting of upwards of two hundred. They begin to assemble about the months of June and July, and generally choose for the place of their future habitation the side of some lake or river. If a lake, in which the water is always pretty nearly of a uniform level, they dispense with building a dam, but if the place they fix upon be the banks of a river, they immediately set about constructing a pier or dam, to confine the water, so that they may always have a good supply."

"That is an instance of very singular sagacity Uncle Thomas. I suppose it is their instinct which teaches them to act in this manner."

"You are right, Frank. Well, the mode in which they set about constructing the dam is this: having fixed upon the spot, they go into the neighbouring forest, and cut quantities of the smaller branches of trees, which they forthwith convey to the place selected, and having fixed them in the earth, interweave them strongly and closely, filling up all the crevices with mud and stones, so as soon to make a most compact construction."

"That must be a work of very great labour, Uncle Thomas."

"The labour is very considerable, Boys; but the power which, for want of a better name, we call Instinct, comes wonderfully to their aid. For instance, it has been observed that they seek all the branches which they want on the banks of the river, higher up than their construction, so that having once got them conveyed to the water, they are easily floated to it."

"Very good, Uncle Thomas."

"When the beavers have finished the dam, they then proceed to construct a house for themselves. First they dig a foundation of greater or less capacity, in proportion to the number of their society. They then form the walls of earth and stones, mixed with billets of wood crossing each other, and thus tying the fabric together just in the same way as you sometimes see masons do in building human dwellings. Their huts are generally of a circular form, something like the figure of a haycock, and they have usually several entrances—one or more opening into the river or lake, below the surface of the water, and one communicating with any bushes and brushwood which may be at hand, so as to afford the means of escape in case of attack either on the land or water side."

"They must be pretty safe then, Uncle Thomas, since they can so readily escape."

"They are pretty secure so long as they have only unreasoning animals to contend with, Frank; but when man, armed with the power, before which mere Instinct must at all times bow, attacks them, they are very easily overcome. Shall I tell you how the hunters capture them?"

"If you please, Uncle Thomas."

"Very well. I must first tell you that the skin of the Beaver is most valuable during winter, as the fur is then thicker and finer than during the summer. They are therefore very little if at all molested during summer by the hunters. When winter sets in, however, and the lakes and rivers are frozen over, a party of hunters set out to seek for the beaver colonies, and, having found them, they make a number of holes in the ice. Having done this and concerted measures, they break down the huts, and the animals instantly get into the water as a place of safety. As they cannot remain long under water, however, they have soon occasion to come to the surface to breathe, and of course make for the holes which the hunters have formed in the ice, when the latter, who are waiting in readiness, knock them on the head."

"But, Uncle Thomas, don't you think it is very cruel to kill the beaver so? I believe it feeds entirely on vegetables, and does no harm to any one."

"You might say the same, John, of the sheep on the downs; the one is not more cruel than the other: both are useful to man, and furnish him with food as well as raiment, and both were, of course, included in the 'dominion' which God originally gave to man 'over the beasts of the field.'"

THE BEAVER.

"Is the beaver used for food, then, Uncle Thomas?"

"It is, and except during a small part of the year, when it feeds on the root of the water-lily, which communicates a peculiar flavour to the flesh of the animal, it is said to be very palatable. It is, however, principally for its fur that it is hunted; the skin, even, is of little value, being coarser and looser in texture, and of course less applicable to general uses, than that of many other animals. I dare say you have often seen it made into gloves.

"I will now read to you an account of a tame beaver, which its owner, Mr. Broderip, communicated to 'the Gardens and Menagerie of the Zoological Society.'

"The animal arrived in this country in the winter of 1825, very young, being small and woolly, and without the covering of long hair, which marks the adult beaver. It was the sole survivor of five or six which were shipped at the same time, and was in a very pitiable condition. Good treatment soon made it familiar. When called by its name, 'Binny,' it generally answered with a little cry, and came to its owner. The hearth rug was its favourite haunt, and thereon it would lie, stretched out, sometimes on its back, and sometimes flat on its belly, but always near its master. The building instinct showed itself immediately after it was let out of its cage, and materials were placed in its way,—and this, before it had been a week in its new quarters. Its strength, even before it was half grown, was great. It would drag along a large sweeping-brush, or a warming-pan, grasping the handle with its teeth, so that the load came over its shoulder, and advancing in an oblique direction, till it arrived at the point where it wished to place it. The long and large materials were always taken first, and two of the longest were generally laid crosswise, with one of the ends of each touching the wall, and the other ends projecting out into the room. The area formed by the crossed brushes and the wall he would fill up with hand-brushes, rush baskets, books, boots, sticks, cloths, dried turf, or any thing portable. As the work grew high, he supported himself on his tail, which propped him up admirably: and he would often, after laying on one of his building materials, sit up over against it, apparently to consider his work, or, as the country people say, 'judge it.' This pause was sometimes followed by changing the position of the material 'judged,' and sometimes it was left in its place. After he had piled up his materials in one part of the room (for he generally chose the same place), he proceeded to wall up the space between the feet of a chest of drawers, which stood at a little distance from it, high enough on its legs to make the bottom a roof for him; using for this purpose dried turf and sticks, which he laid very even, and filling up the interstices with bits of coal, hay, cloth, or any thing he could pick up. This last place he seemed to appropriate for his dwelling; the former work seemed to be intended for a dam. When he had walled up the space between the feet of the chest of drawers, he proceeded to carry in sticks, cloths, hay, cotton, and to make a nest; and, when he had done, he would sit up under the drawers, and comb himself with the nails of his hind feet. In this operation, that which appeared at first to be a malformation, was shown to be a beautiful adaptation to the necessities of the animal. The huge webbed hind feet often turn in, so as to give the appearance of deformities; but if the toes were straight, instead of being incurved, the animal could not use them for the purpose of keeping its fur in order, and cleansing it from dirt and moisture.

"Binny generally carried small and light articles between his right fore leg and his chin, walking on the other three legs; and large masses, which he could not grasp readily with his teeth, he pushed forwards, leaning against them with his right fore paw and his chin. He never carried anything on his tail, which he liked to dip in water, but he was not fond of plunging in his whole body. If his tail was kept moist, he never cared to drink, but, if it was kept dry, it became hot, and the animal appeared distressed, and would drink a great deal. It is not impossible that the tail may have the power of absorbing water, like the skin of frogs, though it must be owned that the scaly integument which invests that member has not much of the character which generally belongs to absorbing surfaces.

" Bread, and bread and milk, and sugar, formed the principal part of Binny's food; but he was very fond of succulent fruits and roots. He was a most entertaining creature; and some highly comic scenes occurred between the worthy, but slow beaver, and a light and airy macauco, that was kept in the same apartment."

" I think I have read, Uncle, that beavers use their tails as trowels to plaster their houses, and as sledges to carry the materials to build huts."

"I dare say, you have, Frank; but I believe such stories are mere fables, told by the ignorant to excite wonder in the minds of the credulous. No such operations have been observed by the most accurate observers of the animal's habits. The wonderful instinct which they display in building their houses is quite sufficient to excite our admiration, without having recourse to false and exaggerated statements."

"The building instinct of the beaver is very curious, Uncle Thomas. Is it displayed by any other animal?"

"All animals exhibit it more or less, Harry, and birds in particular, in the construction of their nests, some of which are very curious indeed; perhaps one of the most striking instances is that of the Sociable Grosbeak, a bird which is found in the interior of the Cape of Good Hope. They construct their nests under one roof, which they

THE SOCIABLE GROSBEAK. 41

form of the branches of some tall and wide-spreading tree, thatching it all over, as it were, with a species of grass.

"When they have got their habitation fairly covered in they lay out the inside, according to some travellers, into regular streets, with nests on both sides, about a couple of inches distant from each other. In one respect, however, they differ from the beaver, they do not appear to lay up a common store of food, the nature of the climate not rendering such a precaution necessary.

"Here is the account of one of these erections furnished by a gentleman who minutely examined the structure.

"I observed on the way a tree with an enormous nest of those birds, to which I have given the appellation of republicans; and, as soon as I arrived at my camp, I despatched a few men, with a waggon, to bring it to me, that I might open the hive, and examine the structure in its minutest parts. When it arrived, I cut it in pieces with a hatchet, and found that the chief portion of the structure consisted of a mass of Boshman's grass, without any mixture, but so compact and firmly basketed together as to be impenetrable to the rain. This is the commencement of the structure; and each bird builds its particular nest under this canopy. But the nests are formed only beneath the eaves of the canopy, the upper surface remaining void, without, however, being useless; for, as it has a projecting rim, and is a little inclined, it serves to let the rain-water run off, and preserves each little dwelling from the rain. Figure to yourself a huge irregular sloping roof, and all the eaves of which are completely covered with nests, crowded one against another, and you will have a tolerably accurate idea of these singular edifices.

"Each individual nest is three or four inches in diameter, which is sufficient for the bird. But as they are all in contact with one another, around the eaves, they appear to the eye to form but one building, and are distinguishable from each other only by a little external aperture, which serves as an entrance to the nest; and even this is sometimes common to three different nests, one of which is situated at the bottom, and the other two at the sides. According to Paterson, the number of cells increasing in proportion to the increase of inhabitants, the old ones become 'streets of communication, formed by line and level.' No doubt, as the republic increases, the cells must be multiplied also; but it is easy to imagine that, as the augmentation can take place only at the surface, the new buildings will necessarily cover the old ones, which must therefore be abandoned.

"Should these even, contrary to all probability, be able to subsist, it may be presumed that the depth of their situation, by preventing any circulation and renewal of the air, would render them so extremely hot as to be uninhabitable. But while they thus become useless, they would remain what they were before, real nests, and change neither into streets nor sleeping-rooms.

"The large nest which I examined was one of the most considerable which I had seen any where on my journey, and contained three hundred and twenty inhabited cells."

"Well, Uncle Thomas, that is very curious; I don't know which most to admire. I rather incline to the beaver however, because of the winter store of food which he lays up."

"There is another animal which displays the building instinct so remarkably, that I must tell you something about it before we part."

"Which is it, Uncle Thomas?"

"It is the white ant of Africa; it is a little animal, scarcely, if at all, exceeding in size those of our own country, yet they construct large nests of a conical or sugar loaf shape, sometimes from ten to twelve feet in height; and one species builds them so strong and compact, that even when they are raised to little more than half their height, the wild-bulls of the country use them as sentinel posts to watch over the safety of the herd which grazes below.

"Mr. Smeathman, a naturalist fully capable to do justice to the nature of these erections, states, that on one occasion he and four men stood on the top of one of them. So you may guess how strong they are."

"Of what are they made, Uncle Thomas? They must be very curious structures. How very different from the ant hills of England!"

"Very different, indeed, John. They are made of clay and sand, and as in such a luxuriant climate they soon become coated over with grass, they quickly assume the appearance of hay-cocks. They are indeed very remarkable structures, whether we consider them externally or internally, and are said to excel those of the beaver and the bee in the same proportion as the inhabitants of the most polished European nation excel the huts of the rude inhabitants of the country where the

STORIES ABOUT INSTINCT.

Termites or white ants abound; while in regard to mere size, Mr. Smeathman calculates that, supposing a man's ordinary height to be six feet, the nests of these creatures may be considered, relative to their size and that of man's, as being raised to four times the height of the largest Egyptian pyramids."

"That is enormous, Uncle Thomas?"

"It is indeed, Frank; but strange though it is, the interior of the nest is even more remarkable, many parts of its construction falling little short of human ingenuity. I need not attempt to describe all its arrangements, which, without a plan, would be nearly unintelligible; but there is one device so admirable that I must point it out to you. The nest is formed of two floors, as it were, and all round the walls are galleries perforated in various winding directions, and leading to the store-houses of the colony, or to the nurseries where the eggs are deposited. As it is sometimes convenient to reach the galleries which open from the upper roof

THE WHITE ANT.

without threading all the intricacies of these winding passages, they construct bridges of a single arch, and thus at once reach the upper roof, from which these diverge. They are thus also saved much labour, in transporting provisions, and in bearing the eggs to the places where they remain till they are hatched."

"That is indeed admirable, Uncle Thomas; they must be very curious animals."

"They are divided into various classes, in the same way as bees; choosing a queen, and some of them acting as workers, &c. But the white ants have a class to which there is nothing similar among any other race of insects. These are what Smeathman calls soldiers, from the duties which they perform. They are much less numerous than the workers, being somewhat in the proportion of one in one hundred. The duty of the soldier-insects is to protect the nest when it is attacked. They are furnished with long and slender jaws, and when enraged bite very fiercely, and sometimes even drive off the negroes who may have attacked them, and even white people suffer severely,—the bite bleeding profusely even through the stocking. Some one who observed the colony alarmed, by having part of the nest broken down, gives the following account of the subsequent operations. One of the soldiers first makes his appearance, as if to see if the enemy be gone, and to learn whence the attack proceeds. By and by two or three others make their appearance, and soon afterwards a numerous body rushes out, which increases in number so long as the attack is continued. They are at this time in a state of the most violent agitation; some employed in beating upon the building with their mandibles, so as to make a noise which may be distinctly heard at the distance of three or four feet. Whenever the attack is discontinued, the soldiers retire first, and are quickly followed by the labourers, which hasten in various directions towards the breach, each with a burden of mortar ready tempered, and thus they soon repair the chasm. Besides the duty of protecting the colony, the soldiers seem to act as overseers of the work, one being generally in attendance on every six or eight hundred; and another, who may be looked upon as commander in chief, takes up his station close to the wall which they are repairing, and frequently repeats the beating which I just mentioned, which is instantly answered by a loud hiss from all the labourers within the dome,—those at work labouring with redoubled energy."

"But, Uncle Thomas, what can be the use of such animals as white ants? I really cannot see what use they are for."

"Well, John, I confess I do not much wonder at your question, though, in putting it, you have forgotten that God makes nothing in vain. Mr. Smeathman, who tells us so much about these curious animals, has answered you by anticipation; and his answer is in such a spirit that I cannot do better than read it to you.

"It may appear surprising how a Being perfectly good should have created animals which seem to serve no other end but to spread destruction and desolation wherever they go. But let us be cautious in suspecting any imperfection in the FATHER OF THE UNIVERSE. What at first sight may seem only productive of mischief, will, upon mature deliberation, be found worthy of that wisdom which planned the most beautiful parts of the world. Many poisons are valuable medicines, Storms are beneficial; and diseases often promote life. These *Termites* are indeed frequently pernicious to mankind, but they are also very useful and even necessary. One valuable purpose which they serve is, to destroy decayed trees and other substances which, if left on the surface of the ground in hot climates, would in a short time pollute the air. In this respect they resemble very much the common flies, which are regarded by mankind in general as noxious and, albeit, as useless beings in creation. But this is certainly for want of consideration. There are not probably in all nature animals of more importance, and it would not be difficult to prove that we should feel the want of one or two large quadrupeds much less than of one or two species of these despicable-looking insects. Mankind in general are sensible that nothing is more disagreeable or more pestiferous than putrid substances; and it is apparent to all who have made observation, that those little insects contribute more to the quick dissolution and dispersion of putrescent matter than any other. They are so necessary in all hot climates, that ever in the open fields a dead animal or small putrid substance cannot be laid upon the ground two minutes before it will be covered with flies and their maggots, which, instantly entering, quickly devour one part, and perforating the rest in various directions, expose the whole to be much sooner decomposed by the elements. Thus it is with the *Termites*. The rapid vegetation in hot climates, of which no idea can be formed by any thing to be seen in this, is equalled by as great a degree of destruction from natural as well as accidental causes. It seems apparent that when anything whatever has arrived at its last degree of perfection, the Creator has decreed that it shall be wholly destroyed as soon as possible, that the face of nature may be speedily adorned with fresh productions in the bloom of spring, or the pride of summer; so when trees and even woods are in part destroyed by tornadoes or fire, it is wonderful to observe how many agents are employed in hastening the total dissolution of the rest. But in hot climates there are none so expert, or who do their business so expeditiously and effectually, as these insects, which in a few weeks destroy and carry away the bodies of large trees, without leaving a particle behind; thus clearing the place for other vegetables which soon fill up every vacancy: and in places where two or three years before there has been a populous town, if the inhabitants, as is frequently the case, have chosen to abandon it, there shall be a very thick wood, and not a vestige of a post to be seen, unless the wood has been of a species which from its hardness is called iron wood."

"Thank you, Uncle Thomas. I see, I was quite wrong in supposing that the ants are of no use. I really did not imagine that they could have been so serviceable."

CHAPTER III.

Uncle Thomas describes the Manner in which Wild Elephants are caught, and relates some Curious Stories of the Cunning, Affection, and Intelligence of the Elephant.

"Well, Boys, you are once more welcome!—I am going to tell you some stories about the Elephant to-night, which I hope will interest you quite as much as those which I told you about the dog. Next to the dog the elephant is one of the most intelligent animals; some of his actions, indeed, seem to be rather the result of reason than mere instinct. But I must first tell you about the animal in its native forests.

"In the luxuriant forests with which a large portion of Asia is covered, this huge animal reigns supreme. Its size and strength easily enable it to overcome the most formidable opponents. The

ELEPHANT HUNTING. 55

intelligence with which it has been endowed by its Creator would make it a most formidable enemy to man, but that the same All-wise Being has graciously endowed it with peaceful and gentle feelings. In its native forests it roams about without seeking to molest any one, and even when caught and tamed it very soon becomes gentle and obedient.

"In the East Indies the elephant is in very general use as a beast of burden. For this purpose it is hunted and caught in great numbers by the Natives, who employ some very ingenious devices to deceive them, and to drive them into the ambuscades which they form for them. The manner in which whole herds are captured is as follows:—

"When the herd is discovered by parties who are sent out for the purpose of reconnoitering, they take notice of the direction in which it is ranging, and as, if their food is plentiful, they generally continue to advance in one direction for miles together, the hunters construct, at a considerable distance in front, a series of enclosures, into which it is their object to drive them.

"When every thing is prepared, the hunters, sometimes to the number of several hundreds, divide themselves into small parties, and form a large circle, so as to surround the herd. Each party generally consists of three men, whose duty it is to light a fire and to clear a footpath between their station and that of their neighbours, so that in this way a communication is kept up by the whole circle, and assistance can at once be afforded at any given point.

"New circles are constantly formed at short distances in advance, so as gradually to drive the animals in the required direction. The hunters are all the while concealed by the luxuriant jungle, and do not show themselves to the elephants at all, but urge them forward by the use of drums, rattles, &c. &c., from the noise of which the animals seek to escape, and thus wander on, feeding as they proceed toward the toils which are prepared for them.

"The *keddah*, or trap, as it may be called, consists of three enclosures, each formed of strong stockades on the outside of deep ditches; the innermost one being the strongest, because by the time they arrive in it, the elephants are generally in a state of great excitement, and would soon break down a fragile enclosure, and make their escape.

"As soon as the herd has entered the first enclosure, strong barricades are erected across the entrance; and as there is no ditch at this point, the hunters take advantage of the remarkable dread which the animal has of fire, to scare them from this most vulnerable part of the fortification. Fires are gradually lit all round the first enclosure, so that the only way of escape which is left is by the entrance to the second.

"At first, as if profiting by their former experience, they generally shun the entrance to the second of the series, but at last, seeing no other chance of escape, the leader of the herd ventures forward, and the rest follow. The gate is instantly shut, and they are in the same manner driven into the third enclosure. Finding no outlet from this they become desperate, scream with tremendous power, and seek to escape by violently attacking the sides of the stockade. At all points, however, they are repulsed by lighted fires, and the tumultuous and exulting shouts of the triumphant hunters.

"In this place of confinement they remain for several days. When their excitement has somewhat subsided, they are enticed one by one to enter a narrow passage leading to the second enclosure. As soon as one enters in, the entrance is closed, and as the passage is so narrow that it cannot turn round, it soon fatigues itself by unavailing exertions to beat down the barrier. Strong ropes with running nooses are now laid down, and no sooner does the animal put his foot within one of them, than the rope is drawn tight by some of the hunters who are stationed on a small scaffold which has been raised over the gateway. In the same manner his other feet are secured. When this has been effected, some of the hunters venture to approach, and tie his hind legs together. Having thus secured him, they can with comparative safety complete their capture. When he is completely secured he is placed between two tame elephants, and led away to the forest and fastened to a tree; and the same operation is repeated, till the whole herd has been secured. At first the rage of the captive is extreme; so long as the animals between which he is led away prisoner remain with him he is comparatively quiet, but when he sees them depart, he is agitated with all the horrors of despair, and makes the most extravagant attempts to regain his liberty. For some time he refuses to eat, but gradually becomes resigned, and feeds freely.

"A keeper is appointed to each animal, as they are secured. His first object is to gain its confidence; supplying it regularly with food, pouring water over its body to keep it cool, and gradually accustoming it to caresses. In the course of five or six weeks he generally obtains a complete ascendency over it; its fetters are removed by degrees, it knows his voice and obeys him, and is then gradually initiated into the objects of its future labours."

"Thank you, Uncle Thomas. I now understand all about elephant-hunting. I could not think how the hunters managed to secure such a huge animal. It seems to be no such difficult task after all."

"It seems easy enough from description, Frank; but it sometimes happens that they break loose, and, irritated by their efforts to escape, they range about in the most furious manner, and as they are very cunning animals, it requires all the circumspection of the hunter to counteract their schemes. I recollect a story which displays this quality in a very strong light.

"During the seige of Bhurtpore, in the year 1805, when the British army, with its countless host of followers and attendants, and thousands of

CUNNING OF THE ELEPHANT. 61

cattle, had been for a long time before the city, the approach of the warm season and of the dry hot winds caused the quantity of water in the neighbourhood of the camps to begin to fail; the ponds or tanks had dried up, and no more water was left than the immense wells of the country could furnish. The multitude of men and cattle that were unceasingly at the wells, occasioned no little struggle for priority in procuring the supply, and the consequent confusion on the spot was frequently very considerable. On one occasion, two elephant-drivers, each with his elephant, the one remarkably large and strong, and the other comparatively small and weak, were at the well together; the small elephant had been provided by his master with a bucket for the occasion, which he carried at the end of his proboscis; but the larger animal being destitute of this necessary vessel, either spontaneously, or by desire of his keeper, seized the bucket, and easily wrested it away from his less powerful fellow-servant. The latter was too sensible of his inferiority openly to resist the insult, though it is obvious that he felt it; and great squabbling and abuse ensued between the keepers.

"At length, the weaker animal, watching the opportunity when the other was standing with his side to the well, retired backwards a few paces, in a very quiet unsuspicious manner, and then rushing forward with all his might, drove his head against the side of the other, and fairly pushed him into the well. It may easily be imagined that great inconvenience was immediately experienced, and serious apprehensions quickly followed, that the water in the well, on which the existence of so many seemed in a great measure to depend, would be spoiled by the unwieldy brute which was precipitated into it; and as the surface of the water was nearly twenty feet below the common level, there did not appear to be any means that could be adopted to get the animal out by main force, without the risk of injuring him. There were many feet of water below the elephant, who floated with ease on its surface, and, experiencing considerable pleasure from his cool retreat, he evinced but little inclination even to exert what means of escape he might himself possess.

"A vast number of fascines (bundles of wood) had been employed by the army in conducting the siege; and at length it occurred to the elephant-keeper, that a sufficient number of these might be lowered into the well, on which the animal might be raised to the top, if it could be instructed as to the necessary means of laying them in regular succession under its feet. Permission having accordingly been obtained from the engineers to use the fascines, the keeper had to teach the elephant the lesson, which, by means of that extraordinary ascendency these men attain over their charge, joined with the intellectual resources of the animal itself, he was soon enabled to do; and the elephant began quickly to place each fascine, as it was lowered, successively under him, until, in a little time, he was enabled to stand upon them. By this time, however, the cunning brute, enjoying the pleasure of his situation, after the heat and partial privation of water to which he had been lately exposed, was unwilling to work any longer; and all the threats of his keeper could not induce him to place another fascine. The man then opposed cunning to cunning, and began to caress and praise the elephant; and what he could not effect by threats he was enabled to do by the repeated promise of plenty of arrack, a spirituous beverage composed of rum, of which the elephant is very fond. Incited by this, the animal again set to work, raised himself considerably higher, until, by a partial removal of the masonry round the top of the well, he was enabled to step out, after having been in the water about fourteen hours."

"That was very cunning, Uncle Thomas. The keepers seem to attain great ascendency over the animals."

"The attachment of the elephant to its keeper, and the command which some of these men acquire

DOCILITY OF THE ELEPHANT. 65

over the objects of their care by appealing to their affections is very extraordinary. The mere sound of the keeper's voice has been known to reclaim an animal which escaped from domestication and resumed its original freedom:—

"A female elephant, belonging to a gentleman in Calcutta, who was ordered from the upper country to Chittagong, in the route thither, broke loose from her keeper, and, making her way to the woods, was lost. The keeper made every excuse to vindicate himself, which the master of the animal would not listen to, but branded the man with dishonesty; for it was instantly supposed that he had sold the elephant. He was tried for it, and condemned to work on the roads for life, and his wife and children sold for slaves.

"About twelve years afterwards, this man, who was known to be well acquainted with breaking elephants, was sent into the country with a party to assist in catching wild ones. They came upon a herd, amongst which the man fancied he saw the long-lost elephant for which he had been condemned. He resolved to approach it, nor could the strongest remonstrances of the party dissuade him from the attempt. As he approached the animal, he called her by name, when she immediately recognised his voice; she waved her trunk in the air as a token of salutation, and kneeling down, allowed him to mount her neck. She afterwards assisted in taking other elephants, and decoyed three young ones, to which she had given birth since her escape. The keeper returned to his master, and the singular circumstances attending the recovery of the elephant being told, he regained his character; and, as a recompense for his unmerited sufferings, had a pension settled on him for life."

"That was an instance of rare good fortune, Uncle Thomas. How very curious that he should fall in with the herd in which his own elephant was!"

"It was very fortunate indeed, Frank. It was

DOCILITY OF THE ELEPHANT.

not a little curious too that the elephant should recognise him after so long a period. But the attachment which they show to their keepers is sometimes very great. One which in a moment of rage killed its keeper a few years ago, adopted his son as its *carnac* or driver, and would allow no one else to assume his place. The wife of the unfortunate man was witness to the dreadful scene, and, in the frenzy of her mental agony, took her two children, and threw them at the feet of the elephant, saying, 'As you have slain my husband, take my life also, as well as that of my children!' The elephant became calm, seemed to relent, and as if stung with remorse, took up the eldest boy with its trunk, placed him on its neck, adopted him for its carnac, and never afterwards allowed another to occupy that seat."

"That was at least making all the reparation in its power, Uncle Thomas."

"There is one or two other stories about the elephant, showing that he knows how to revenge an insult, which I must tell you before you go.

"A merchant at Bencoolen kept a tame elephant, which was so exceedingly gentle in his habits, that he was permitted to go at large. This huge animal used to walk about the streets in the most quiet and orderly manner, and paid many visits through the city to people who were kind to him. Two cobblers took an ill will to this inoffensive creature, and several times pricked him on the proboscis with their awls. The noble animal did not chastise them in the manner he might have done, and seemed to think they were too contemptible to be angry with them. But he took other means to punish them for their cruelty. He filled his trunk with water of a dirty quality, and advancing towards them in his ordinary manner, spouted the whole of the puddle over them. The punishment was highly applauded by those who witnessed it, and the poor cobblers were laughed at for their pains."

THE ELEPHANT AND COBBLERS. Page 68.

"Ha! ha! ha! He must have been a very knowing animal, Uncle Thomas. I dare say, the cobblers behaved better in future."

"I dare say they would, Boys. Here is another story of the same description, but the trickster did not escape so easily."

"A person in the island of Ceylon, who lived near a place where elephants were daily led to water, and often sat at the door of his house, used occasionally to give one of these animals some fig leaves, a food to which elephants are very partial. Once he took it into his head to play one of the elephants a trick. He wrapped a stone round with fig leaves, and said to the carnac, 'This time I will give him a stone to eat, and see how it will agree with him.' The carnac answered, 'that the elephant would not be such a fool as to swallow a stone.' The man, however, reached the stone to the elephant, who, taking it with his trunk, immediately let it fall to the ground. 'You see,' said the keeper, 'that I was right;' and without further words, drove away his elephants. After they were watered, he was conducting them again to their stable. The man who had played the elephant the trick was still sitting at his door, when, before he was aware, the animal ran at him, threw his trunk around his body, and, dashing him to the ground, trampled him immediately to death."

CHAPTER IV.

Uncle Thomas introduces to the Notice of the Young Folks the Ettrick Shepherd's Stories about Sheep; and tells them some Interesting Stories about the Goat, and its Peculiarities.

"I DARE say, Boys, you have not forgotten the Ettrick Shepherd's wonderful stories about his dogs. Some of those which he relates about sheep are equally remarkable, and as he tells them in the same pleasing style, I think I cannot do better than read to you the chapter in 'The Shepherd's Calendar' which he devotes to this animal."

"Thank you, Uncle Thomas. We remember very well his stories about Sirrah and Hector and Chieftain, and the old Shepherd's grief at parting with his dog."

"That's right, Boys; I am pleased to think that you do not forget what I tell you. But listen to the Ettrick Shepherd."

"The sheep has scarcely any marked character save that of natural affection, of which it possesses a very great share. It is otherwise a stupid indifferent animal, having few wants, and fewer expedients. The old black-faced, or forest breed, have far more powerful capabilities than any of the finer breeds that have been introduced into Scotland, and, therefore, the few anecdotes that I have to relate shall be confined to them.

"So strong is the attachment of the sheep to the place where they have been bred, that I have heard of their returning from Yorkshire to the Highlands. I was always somewhat inclined to suspect that they might have been lost by the way, but it is certain, however, that when once one or a few sheep get away from the rest of their acquaintances, they return homeward with great eagerness and perseverance. I have lived beside a drove-road the better part of my life, and many stragglers have I seen bending their steps northward in the spring of the year. A shepherd rarely sees these journey-

LOVE OF HOME. 73

ers twice; if he sees them, and stops them in the morning, they are gone long before night; and if he sees them at night, they will be gone many miles before morning. This strong attachment to the place of their nativity is much more predominant in our old aboriginal breed than in any of the other kinds with which I am acquainted.

" The most singular instance that I know of, to be quite well authenticated, is that of a black ewe, that returned with her lamb from a farm in the head of Glen-Lyon, to the farm of Harehope, in Tweeddale, and accomplished the journey in nine days. She was soon missed by her owner, and a shepherd was despatched in pursuit of her, who followed her all the way to Crieff, where he turned, and gave her up. He got intelligence of her all the way, and every one told him that she absolutely persisted in travelling on,—she would not be turned, regarding neither sheep nor shepherd by the way. Her lamb was often far behind, and she had constantly to urge it on by impatient bleating. She unluckily came to Stirling on the morning of a great annual fair, about the end of May, and judging it imprudent to venture through the crowd with her lamb, she halted on the north side of the town the whole day, where she was seen by hundreds, lying close by the road-side. But next morning, when all became quiet, a little after the break of day, she was observed stealing quietly through the town, in apparent terror of the dogs that were prowling about the street. The last time she was seen on the road was at a toll-bar near St. Ninian's; the man stopped her, thinking she was a strayed animal, and that some one would claim her. She tried several times to break through by force when he opened the gate, but he always prevented her, and at length she turned patiently back. She had found some means of eluding him, however, for home she came on a Sabbath morning, early in June; and she left the farm of Lochs, in Glen-Lyon, either on the Thursday afternoon, or Friday morning, a week and two days before. The farmer

NATURAL AFFECTION. 75

of Harehope paid the Highland farmer the price of her, and she remained on her native farm till she died of old age, in her seventeenth year.

"With regard to the natural affection of this animal, the instances that might be mentioned are without number. When one loses its sight in a flock of sheep, it is rarely abandoned to itself in that hapless and helpless state. Some one always attaches itself to it, and by bleating calls it back from the precipice, the lake, the pool, and all dangers whatever. There is a disease among sheep, called by shepherds the Breakshugh, a deadly sort of dysentery, which is as infectious as fire, in a flock. Whenever a sheep feels itself seized by this, it instantly withdraws from all the rest, shunning their society with the greatest care; it even hides itself, and is often very hard to be found. Though this propensity can hardly be attributed to natural instinct. it is, at all events, a provision of nature of the greatest kindness and beneficence.

STORIES ABOUT INSTINCT.

"Another manifest provision of nature with regard to these animals is, that the more inhospitable the land is on which they feed, the greater their kindness and attention to their young. I once herded two years on a wild and bare farm called Willenslee, on the border of Mid-Lothian, and of all the sheep I ever saw, these were the kindest and most affectionate to their lambs. I was often deeply affected at scenes which I witnessed. We had one very hard winter, so that our sheep grew lean in the spring, and the thwarter-ill (a sort of paralytic affection) came among them, and carried off a number. Often have I seen these poor victims, when fallen down to rise no more, even when unable to lift their heads from the ground, holding up the leg, to invite the starving lamb to the miserable pittance that the udder still could supply. I had never seen aught more painfully affecting.

"It is well known that it is a custom with shepherds, when a lamb dies, if the mother have a sufficiency of milk, to bring her from the hill, and put another lamb to her. This is done by putting the skin of the dead lamb upon the living one; the ewe immediately acknowledges the relationship, and after the skin has warmed on it, so as to give it something of the smell of her own progeny, and it has sucked her two or three times, she accepts and nourishes it as her own ever after. Whether it is from joy at this apparent reanimation of her young one, or because a little doubt remains on her mind which she would fain dispel, I cannot decide; but, for a number of days, she shows far more fondness, by bleating and caressing over this one, than she did formerly over the one that was really her own. But this is not what I wanted to explain; it was, that such sheep as thus lose their lambs must be driven to a house with dogs, so that the lamb may be put to them; for they will only take it in a dark confined place. But at Willenslee, I never needed to drive home a sheep by force, with dogs, or in any other way than the following:

I found every ewe, of course, standing hanging her head over her dead lamb; and having a piece of twine with me for the purpose, I tied that to the lamb's neck or foot, and trailing it along, the ewe followed me into any house or fold that I choose to lead her. Any of them would have followed me in that way for miles, with her nose close on the lamb, which she never quitted for a moment, except to chase my dog, which she would not suffer to walk near me. I often, out of curiosity, led them in to the side of the kitchen fire by this means, into the midst of servants and dogs; but the more that dangers multiplied around the ewe, she clung the closer to her dead offspring, and thought of nothing whatever but protecting it. One of the two years, while I remained on this farm, a severe blast of snow came on by night, about the latter end of April, which destroyed several scores of our lambs; and as we had not enow of twins and odd lambs for the mothers that had lost theirs, of course we selected the best ewes, and put lambs to them. As we were making the distribution, I requested of my master to spare me a lamb for a hawked ewe which he knew, and which was standing over a dead lamb in the head of the Hope, about four miles from the house. He would not do it, but bid me let her stand over her lamb for a day or two, and perhaps a twin would be forthcoming. I did so, and faithfully she did stand to her charge; so faithfully, that I think the like never was equalled by any of the woolly race. I visited her every morning and evening, and for the first eight days never found her above two or three yards from the lamb; and always, as I went my rounds, she eyed me long ere I came near her, and kept tramping with her feet, and whistling through her nose, to frighten away the dog; he got a regular chase twice a day as I passed by: but, however excited and fierce a ewe may be, she never offers any resistance to mankind, being perfectly and meekly passive to them. The weather grew fine and warm, and the dead lamb soon decayed, which the body of a dead lamb does particularly soon: but still this affectionate and desolate creature kept hanging over the poor remains with an attachment that seemed to be nourished by hopelessness. It often drew the tears from my eyes to see her hanging with such fondness over a few bones, mixed with a small portion of wool. For the first fortnight she never quitted the spot, and for another week she visited it every morning and evening, uttering a few kindly and heart-piercing bleats each time; till at length every remnant of her offspring vanished, mixing with the soil, or wafted away by the winds."

"Poor creature! Uncle Thomas, that was very affecting."

"So much for the Ettrick Shepherd. I will now tell you a story about a remarkable instance of sagacity in a sheep, of which I myself was an eye-witness.

"One evening, as I was enjoying a walk through some verdant pastures, which were plentifully dotted with sheep, my attention was attracted by

THE RESCUED LAMB.

the motions of one which repeatedly came close up to me, bleating in a piteous manner, and after looking expressively in my face, ran off towards a brook which meandered through the midst of the pastures. At first I took little notice of the creature, but as her entreaties became importunate, I followed her. Delighted at having at length attracted my notice, she ran with all her speed, frequently looking back. When I reached the spot, I discovered the cause of all her anxiety; her lamb had unfortunately fallen into the brook, whose steep banks prevented it from making its escape. Fortunately the water, though up to the little creature's back, was not sufficient to drown it. I rescued it with much pleasure, and to the great gratification of its affectionate mother, who licked it with her tongue to dry it, now and then skipping about, and giving vent to her joy and gratitude in most expressive gambols.

"Though differing in many respects from the sheep, the goat bears so strong a resemblance to that animal, that, now that I am speaking of it, I may as well tell you a story or two about the goat. It will save my returning to it afterwards."

"Very well, Uncle Thomas."

"The goat is in every respect more fitted for a life of savage liberty than the sheep. It is of a more lively disposition, and is possessed of a greater degree of instinct. It readily attaches itself to man, and seems sensible of his caresses. It delights in climbing precipices, and going to the very edge of danger, and it is often seen suspended upon an eminence overhanging the sea, upon a very little base, and sometimes even sleeps there in security. Nature has in some measure fitted it for traversing these declivities with ease; the hoof is hollow underneath, with sharp edges, so that it walks as securely on the ridge of a house as on the level ground.

"When once reduced to a state of domestication, the goat seldom resumes its original wildness. A good many years ago, an English vessel happening to touch at the island of Bonavista, two negroes

THE ARAB AND HIS GOAT. 83

came and offered the sailors as many goats as they chose to take away. Upon the captain expressing his surprise at this offer, the negroes assured him that there were but twelve persons on the island, and that the goats had multiplied in such a manner as even to become a nuisance: they added, that far from giving any trouble to capture them, they followed the few inhabitants that were left with a sort of obstinacy, and became even troublesome by their tameness. The celebrated traveller Dr. Clarke gives a very curious account of a goat, which was trained to exhibit various amusing feats of dexterity.

"We met, (says he,) an Arab with a goat which he led about the country to exhibit, in order to gain a livelihood for itself and its owner. He had taught this animal, while he accompanied its movements with a song, to mount upon little cylindrical blocks of wood, placed successively one above another, and in shape resembling the dice-box belonging to a backgammon table. In this manner the goat stood, first, on the top of two; afterwards, of three, four, five, and six, until it remained balanced upon the summit of them all, elevated several feet above the ground, and with its four feet collected upon a single point, without throwing down the disjointed fabric on which it stood. The diameter of the upper cylinder, on which its four feet alternately remained until the Arab had ended his ditty, was only two inches, and the length of each six inches. The most curious part of the performance occurred afterwards; for the Arab, to convince us of the animal's attention to the turn of the air, interrupted the *Da Capo;* and, as often as he did this, the goat tottered, appeared uneasy, and, upon his becoming suddenly silent, in the middle of his song, it fell to the ground.

"Like the sheep, the goat possesses great natural affection for its young. In its defence it boldly repels the attacks of the most formidable opponents. I remember a little story which finely illustrates this instinctive courage.

THE ARAB AND HIS GOAT. Page 84.

THE GOAT.

"A person having missed one of his goats when his flock was taken home at night, being afraid the wanderer would get among the young trees in his nursery, two boys, wrapped in their plaids, were ordered to watch all night. The morning had but faintly dawned, when they set out in search of her. They at length discovered her on a pointed rock at a considerable distance, and hastening to the spot perceived her standing watching her kid with the greatest anxiety, and defending it from a fox. The enemy turned round and round to lay hold of his prey, but the goat presented her horns in every direction. The youngest boy was despatched for assistance to attack the fox, and the eldest, hallooing and throwing up stones, sought to intimidate it as he climbed to rescue his charge. The fox seemed well aware that the child could not execute his threats; he looked at him one instant, and then renewed the assault, till, quite impatient, he made a sudden effort to seize the kid. The whole three suddenly disappeared, and were found at the bottom of the precipice. The goat's horns were darted into the back of the fox; the kid lay stretched beside her. It is supposed that the fox had fixed his teeth in the kid, for its neck was lacerated; but that when the faithful mother inflicted a death wound upon her mortal enemy he probably staggered, and brought his victims with him over the rock.

"There is another story of the goat, which places its gratitude and affection in such an interesting light, that I am sure it will delight you:—

"After the final suppression of the Scottish Rebellion of 1715, by the decisive Battle of Preston, a gentleman who had taken a very active share in it escaped to the West Highlands to the residence of a female relative, who afforded him an asylum. As in consequence of the strict search which was made after the ringleaders, it was soon judged unsafe for him to remain in the house of his friend, he was conducted to a cavern in a sequestered situation, and furnished with a supply of food. The approach to this lonely abode consisted of a

THE GRATEFUL GOAT.

small aperture, through which he crept, dragging his provisions along with him. A little way from the mouth of the cave the roof became elevated, but on advancing, an obstacle obstructed his progress. He soon perceived that, whatever it might be, the object was a living one, but unwilling to strike at a venture with his dirk, he stooped down, and discovered a goat and her kid lying on the ground. The animal was evidently in great pain, and feeling her body and limbs, he ascertained that one of her legs had been fractured. He bound it up with his garter, and offered her some of his bread; but she refused to eat, and stretched out her tongue, as if intimating that her mouth was parched with thirst. He gave her water, which she drank greedily, and then she ate the bread. At midnight he ventured from the cave, pulled a quantity of grass and the tender branches of trees, and carried them to the poor sufferer, which received them with demonstrations of gratitude.

"The only thing which this fugitive had to

arrest his attention in his dreary abode, was administering comfort to the goat; and he was indeed thankful to have any living creature beside him. It quickly recovered, and became tenderly attached to him. It happened that the servant who was intrusted with the secret of his retreat fell sick, when it became necessary to send another with provisions. The goat, on this occasion, happening to be lying near the mouth of the cavern, opposed his entrance with all her might, butting him furiously; the fugitive, hearing a disturbance, went forward, and receiving the watchword from his new attendant, interposed, and the faithful goat permitted him to pass. So resolute was the animal on this occasion, that the gentleman was convinced she would have died in his defence."

CHAPTER V.

Uncle Thomas relates some Very Remarkable Stories about the Cat; points out to the Boys the Connexion subsisting between the Domestic Cat and the Lion, Tiger, &c., and tells them some Stories about the Gentleness, as well as the Ferocity of these Animals.

" Though far from being so general a favourite as the dog, the domestic cat has many qualities to recommend it to attention and regard, and some of the stories which I am going to tell you exhibit instances of instinctive attachment and gentleness which cannot be surpassed.

" Here is one of attachment, which will match with the best of those of the dog.

" A cat which had been brought up in a family became extremely attached to the eldest child, a little boy, who was very fond of playing with her. She bore with the most exemplary patience any maltreatment which she received from him—which even good-natured children seldom fail, occasionally, to give to animals in their sports with them—without ever making any attempt at resistance. As the cat grew up, however, she daily quitted her playfellow for a time, from whom she had formerly been inseparable, in order to follow her natural propensity to catch mice; but even when engaged in this employment, she did not forget her friend; for, as soon as she had caught a mouse, she brought it alive to him. If he showed an inclination to take her prey from her, she anticipated him, by letting it run, and waited to see whether he was able to catch it. If he did not, the cat darted at, seized it, and laid it again before him; and in this manner the sport continued as long as the child showed any inclination for the amusement.

"At length the boy was attacked by small-pox, and, during the early stages of his disorder, the cat never quitted his bed-side; but, as his

FELINE AFFECTION. 91

danger increased, it was found necessary to remove the cat and lock it up. The child died. On the following day, the cat having escaped from her confinement, immediately ran to the apartment where she hoped to find her playmate. Disappointed in her expectation, she sought for him with symptoms of great uneasiness and loud lamentation, all over the house, till she came to the door of the room in which the corpse lay. Here she lay down in silent melancholy, till she was again locked up. As soon as the child was interred, and the cat set at liberty, she disappeared; and it was not till a fortnight after that event, that she returned to the well-known apartment, quite emaciated. She would not, however, take any nourishment, and soon ran away again with dismal cries. At length, compelled by hunger, she made her appearance every day at dinner-time, but always left the house as soon as she had eaten the food that was given her. No one knew where she spent the rest of her time, till she was found one day under the wall of the burying-ground, close to the grave of her favourite; and so indelible was the attachment of the cat to her deceased friend, that till his parents removed to another place, five years afterwards, she never, except in the greatest severity of winter, passed the night any where else than at the above-mentioned spot, close to the grave. Ever afterwards she was treated with the utmost kindness by every person in the family. She suffered herself to be played with by the younger children, although without exhibiting a particular partiality for any of them.

"There is another story of the cat's attachment, of a somewhat less melancholy cast, which I lately saw recorded in a provincial newspaper.

"A country gentleman of our acquaintance, who is neither a friend to thieves nor poachers, has at this moment in his household a favourite cat, whose honesty, he is sorry to say, there is but too

THE AFFECTIONATE CAT. Page 92.

THE PHILANTHROPIC CAT. 93

much reason to call in question. The animal, however, is far from being selfish in her principles; for her acceptable gleanings she regularly shares among the children of the family in which her lot is cast. It is the habit of grimalkin to leave the kitchen or parlour, as often as hunger and an opportunity may occur, and wend her way to a certain pastrycook's shop, where, the better to conceal her purpose, she endeavours slily to ingratiate herself into favour with the mistress of the house. As soon as the shopkeeper's attention becomes engrossed in business, or otherwise, puss contrives to pilfer a small pie or tart from the shelves on which they are placed, speedily afterwards making the best of her way home with her booty. She then carefully delivers her prize to some of the little ones in the nursery. A division of the stolen property quickly takes place; and here it is singularly amusing to observe the cunning animal, not the least conspicuous among the numerous group, thankfully mumping her share of the illegal traffic. We may add that the pastrycook is by no means disposed to institute a legal process against the delinquent, as the children of the gentleman to whom we allude are honest enough to acknowledge their four-footed playmate's failings to papa, who willingly compensates any damage the pastrycook may sustain from the petty depredations of the would-be philanthropic cat.

"I remember how highly pleased you were with the story which I told you about the dog discovering the murderers of his master. There is one of a very similar description of a French cat, which I am sure will equally interest you.

"In the beginning of the present century a woman was murdered in Paris. The magistrate who went to investigate the affair was accompanied by a physician; they found the body lying upon the floor, and a greyhound watching over it, and howling mournfully. When the gentleman entered the apartment, it ran to them without barking, and then returned with a melancholy mien to the body of his murdered mistress. Upon a chest in a corner of the room sat a cat, motionless, with eyes expressive of furious indignation, stedfastly fixed upon the body. Many persons now entered the apartment, but neither the appearance of such a crowd of strangers, nor the confusion that prevailed in the place, could make her change her position. In the mean time, some persons were apprehended on suspicion of being the murderers, and it was resolved to lead them into the apartment. Before the cat got sight of them, when she only heard their footsteps approaching, her eyes flashed with increased fury, her hair stood erect, and so soon as she saw them enter the apartment, she sprang towards them with expressions of the most violent rage, but did not venture to attack them, being probably alarmed by the numbers that followed. Having turned several times towards them with a peculiar ferocity of aspect, she crept into a corner, with an air indicative of the deepest melancholy. This behaviour of the cat astonished every one present. The effect which it produced upon the murderers was such as almost to amount to an acknowledgment of guilt. Nor did this remain long doubtful, for a train of accessory circumstances was soon discovered which proved it to complete conviction.

"I have often warned you against stories of ghosts and hobgoblins, and shown you on how frail a foundation they generally rest. There is a story in which a cat was one of the principal actors, which contains the elements of as marvellous a tale of this description as could be desired. It happened in the west of Scotland.

"Some years ago, a poor man whose habits of life had always been of the most retired description, giving way to the natural despondency of his disposition, put an end to his existence. The only other inmate of his cottage was a favourite cat. When the deed was discovered, the cat was found assiduously watching over her late master's body, and it was with some difficulty she could be driven away.

"The appalling deed naturally excited a great deal of attention in the surrounding neighbourhood; and on the day after the body was deposited in the grave, which was made at the outside of the church-yard, a number of school-boys ventured thither, to view the resting-place of one who had at times been the subject of village wonder, and whose recent act of self-destruction was invested with additional interest. At first, no one was brave enough to venture near; but at last, the appearance of a hole in the side of the grave irresistibly attracted their attention. Having been minutely examined, it was at length determined that it must have been the work of some body-snatcher, and the story having spread, the grave was minutely examined, but as the body had not been removed, the community considered themselves fortunate in having made so narrow an escape. The turf was replaced, and the grave again carefully covered up.

"On the following morning the turf was again displaced, and a hole, deeper than before, yawned in the side of the sad receptacle. Speculation was soon busy at work, and all sorts of explanations were suggested. In the midst of their speculations, alarmed perhaps by the noise of the disputants, poor puss darted from the hole, much to the confusion of some of the most noisy and dogmatic expounders of the mystery. Again the turf was replaced, and again and again was it removed by the unceasing efforts of the faithful cat to share the resting-place of her deceased master. It was at last found necessary to shoot her, it being found impossible otherwise to put a stop to her unceasing importunities."

"Poor puss! What a pity it should have been necessary to destroy such a faithful animal. I wonder no one tried to gain its affections, and thus charm it from its dreary abode. Uncle Thomas, did you ever hear Dr. Good's account of a very extraordinary instance of sagacity exemplified by his cat? I was very much struck with it when I saw it a day or two ago in his 'Book of Nature.' If you please, I will read it to you."

"Very well, Harry, I shall be glad to hear it; I dare say it is an old acquaintance of mine. I have been such a diligent searcher after stories of this description, that I think very few have escaped me."

"A favourite cat, that was accustomed from day to day to take her station quietly at my elbow, on the writing table, sometimes for hour after hour, whilst I was engaged in study, became at length less constant in her attendance, as she had a kitten to take care of. One morning she placed herself in the same spot, but seemed unquiet, and, instead of seating herself as usual, continued to rub her furry sides against my hand and pen, as though resolved to draw my attention, and make me leave off. As soon as she had accomplished this point, she leaped down on the carpet, and made towards the door, with a look of great uneasiness. I opened the door for her, as she seemed to desire, but, instead of going forward, she turned round, and looked earnestly at me, as though she wished me to follow her, or had something to communicate. I did not fully understand her meaning, and, being much engaged at the time, shut the door upon her, that she might go where she liked.

"In less than an hour afterwards, however, she had again found an entrance into the room, and drawn close to me, but, instead of mounting the table, and rubbing herself against my hand, as before, she was now under the table, and continued to rub herself against my feet, on moving which I struck them against a something which seemed to be in their way, and, on looking down, beheld with equal grief and astonishment the dead body of her little kitten which I supposed had been alive and in good health, covered over with cinder dust. I now entered into the entire train of this afflicted cat's feelings. She had suddenly lost the nursling she doated on, and was resolved to make me acquainted with it,—assuredly that I might know her grief, and probably also that I might inquire into the cause, and, finding me too dull to understand her expressive motioning that I would follow her to the cinder heap, on which the dead kitten had been thrown, she took the great labour of bringing it to me herself, from the area on the basement floor, and up a whole flight of stairs, and laid it at my feet. I took up the kitten in my hand, the cat still following me, made inquiry into the cause of its death, which I found, upon summoning the servants, to have been an accident, in which no one was much to blame; and the yearning mother having thus obtained her object, and gotten her master to enter into her cause, and divide her sorrows with her, gradually took comfort, and resumed her former station by my side."

"Thank you, Harry, I do not think I ever heard that story before. Here is one that will match it however, displaying considerable ingenuity in a cat in the protection of her young.

"A cat belonging to Mr. Stevens, of the Red Lion Hotel, Truro, having been removed from that town to a barn at some distance, soon afterwards produced four kittens. Not wishing the stock increased, Mr. Stevens desired three of them to be drowned, next morning, before opening their eyes on the world. Puss was deeply affected by this bereavement, and resolved on moving her remaining offspring to a place of security. When the person appointed to feed grimalkin went with her breakfast next day, no traces of her or her kitten were to be found. He called; but all was silent as the tomb; every corner was searched in vain; no cat was forthcoming. Here the matter rested for several days, when, at length, early one morning, puss made her appearance in the court of her master's house, a melancholy picture of starvation. Having satisfied her hunger, and loitered about the house during the day, late in the evening she took her departure, carrying away some meat. For several days she continued her visits in the same manner, taking care never to leave home empty-mouthed at night. Her proceedings having excited attention, she was followed by two men, in one of her nocturnal retreats, and traced to the top of a wheat stack, at some distance. On obtaining a ladder, her surviving kitten was found, in a curiously constructed hole, sleek and plump, but as wild as a young tiger, and would allow no one to touch it. A few days afterwards, the mother finding, perhaps, that her own daily journeys were rather fatiguing, or thinking it was time that the object of her solicitude should be introduced into the world, or, probably, that the kitten had attained an age when it could protect itself, she took advantage of a dark and silent night, when cat-worrying dogs and boys were reposing, to convey it safely to Truro, where tabby and her kitten found a welcome reception.

"Though from bad education the cat and dog

are generally the most determined enemies, some instances have occurred of the greatest friendship subsisting between these animals. Here is an instance recorded by a French author on the Language of Brutes.

"I had a cat and dog, which became so attached to each other, that they would never willingly be asunder. Whenever the dog got any choice morsel of food, he was sure to divide it with his whiskered friend. They always ate sociably out of one plate, slept in the same bed, and daily walked out together. Wishing to put this apparently sincere friendship to the proof, I, one day, took the cat by herself into my room, while I had the dog guarded in another apartment. I entertained the cat in a most sumptuous manner, being desirous to see what sort of a meal she would make without her friend, who had hitherto been her constant table companion. The cat enjoyed the treat with great glee, and seemed to have entirely forgotten the dog. I had had a partridge for dinner, half of which I intended to keep for supper. My wife covered it with a plate, and put it into a cupboard, the door of which she did not lock. The cat left the room, and I walked out upon business. My wife, meanwhile, sat at work in an adjoining apartment. When I returned home, she related to me the following circumstances :—The cat, having hastily left the dining room, went to the dog, and mewed uncommonly loud, and in different tones of voice; which the dog, from time to time, answered with a short bark. They both then went to the door of the room where the cat had dined, and waited till it was opened. One of my children opened the door, and immediately the two friends entered the apartment. The mewing of the cat excited my wife's attention. She rose from her seat, and stepped softly up to the door, which stood ajar, to observe what was going on. The cat led the dog to the cupboard which contained the partridge, pushed off the plate which covered it, and, taking out my intended supper, laid it before her canine friend, who devoured it greedily. Probably the cat, by her mewing, had given the dog to understand what an excellent meal she had made, and how sorry she was that he had not participated in it; but, at the same time, had explained to him that something was left for him in the cupboard, and persuaded him to follow her thither. Since that time I have paid particular attention to these animals, and am perfectly convinced that they communicate to each other whatever seems interesting."

"Oh! indeed, Uncle Thomas, do you think that animals understand each other?"

"I have no doubt that they do to a limited extent, Harry, but I cannot go the whole length of Monsieur Wenzel, who records the story I have just told you.

"I will now tell you some stories about some of the other animals of the cat kind, such as the lion, tiger, &c.; and though these animals differ so much from the domestic cat, they all belong to the same family; the huge lion, which carries off with ease a buffalo from the herd, or makes the forest tremble with his hoarse roar is no more than an enormous cat.

"I dare say you have all heard the story of 'Androcles and the Lion,' which is recorded in that most delightful book, 'Sandford and Merton.' It is so captivating a tale, that I must repeat it to you as much for my own gratification as for yours. I will just observe, however, that it is a fiction, and not a real story, though I can tell you one or two very similar ones, which occurred in real life."

"There was a certain slave named Androcles, who was so ill treated by his master that his life became insupportable. Finding no remedy from what he suffered, he at length said to himself:—'It is better to die than to continue to live in such hardships and misery as I am obliged to suffer. I am determined, therefore, to run away from my master; if I am taken again, I know that I shall be punished with a cruel death, but it is better to die at once, than to live in misery. If I escape, I must betake myself to deserts and woods, inhabited only by wild beasts, but they cannot use me more cruelly than I have been by my fellow-creatures, therefore I will rather trust myself to them, than continue to be a miserable slave.

"Having formed this resolution, he took an opportunity of leaving his master's house, and hid himself in a thick forest, which was some miles distant from the city. But here the unhappy man found that he had only escaped from one kind of misery to experience another. He wandered about all day through a vast and trackless wood, where his flesh was continually torn by thorns and brambles. He grew hungry, but he could find no food in this dreary solitude. At length he was ready to die with fatigue, and lay down in despair in a large cavern.

"The unfortunate man had not been long quiet in the cavern, before he heard a dreadful noise, which seemed to be the roar of some wild beast, and terrified him very much. He started up with a design to escape, and had already reached the mouth of the cave, when he saw coming towards him a lion of prodigious size, which prevented any possibility of retreat. He now believed his destruction to be inevitable, but to his great astonishment the beast advanced towards him with a gentle pace, without any mark of enmity or rage, and uttered a kind of mournful voice, as if he demanded the assistance of the man.

"Androcles, who was naturally of a resolute disposition, acquired courage from this circumstance to examine his monstrous guest, who gave him sufficient leisure for this purpose. He saw, as the lion approached him, that he seemed to limp upon one of his legs, and that the foot was extremely swelled, as if it had been wounded. Acquiring still more fortitude from the gentle demeanour of the beast, he advanced towards him, and took hold of the wounded part as a surgeon would examine his patient. He then perceived that a thorn of uncommon size had penetrated the ball of the foot, and was the occasion of the swelling and the lameness which he had observed. Androcles found that the beast, far from resenting his familiarity, received it with the greatest gentleness, and seemed to invite him by his blandishments to proceed. He therefore extracted the thorn, and, pressing the swelling, discharged a considerable quantity of matter, which had been the cause of so much pain. As soon as the beast felt himself thus relieved, he began to testify his joy and gratitude by every expression in his power. He jumped about like a wanton spaniel, wagged his enormous tail, and licked the feet and hands of his physician. Nor was he contented with these demonstrations of kindness. From this moment Androcles became his guest; nor did the lion ever sally forth in quest of his prey, without bringing home the produce of his chase, and sharing it with his friend.

"In this savage state of hospitality did the man

ANDROCLES AND THE LION. Page 110.

continue to live during several months. At length, wandering unguardedly through the woods, he met with a company of soldiers sent out to apprehend him, and was by them taken prisoner, and conducted back to his master. The laws of that country being very severe against slaves, he was tried and found guilty of having fled from his master, and as a punishment for his pretended crime, he was sentenced to be torn in pieces by a furious lion, kept many days without food, to inspire him with additional rage.

"When the destined moment arrived, the unhappy man was exposed, unarmed, in the middle of a spacious arena, inclosed on every side, round which many thousand people were assembled to view the mournful spectacle. Presently a dreadful yell was heard, which struck the spectators with horror, and a monstrous lion rushed out of a den, which was purposely set open, with erected mane and flaming eyes, and jaws that gaped like an open sepulchre. A mournful silence instantly prevailed.

All eyes were turned upon the destined victim, whose destruction seemed inevitable. But the pity of the multitude was soon converted into astonishment, when they beheld the lion, instead of destroying its defenceless enemy, crouch submissively at his feet, fawn upon him as a faithful dog would do upon his master, and rejoice over him as a mother that unexpectedly recovers her offspring. The governor of the town, who was present, then called out with a loud voice, and ordered Androcles to explain to them this unintelligible mystery, and how a savage of the fiercest and most unpitying nature should thus in a moment have forgotten his innate disposition, and be converted into a harmless and inoffensive animal. Androcles then related to the assembly every circumstance of his adventures, and concluded by saying, that the very lion which now stood before them, had been his friend and entertainer in the woods. All present were astonished and delighted with the story, to find that even the fiercest beasts are capable of being softened by gratitude; and, being moved by humanity, they unanimously joined to entreat for the pardon of the unhappy man, from the governor of the place. This was immediately granted to him, and he was also presented with the lion, which had twice saved the life of Androcles."

"Oh, what a delightful story, Uncle Thomas! What a pity it is that it is not true."

"I can tell you one which is true, John which is hardly, if at all, inferior in interest:—

"Sir George Davis, who was English consul at Naples about the middle of the seventeenth century, happening on one ocassion to be in Florence, visited the Menagerie of the Grand Duke. At the farther end of one of the dens he saw a lion which lay in sullen majesty, and which the keepers informed him they had been unable to tame, although every effort had been used for upwards of three years. Sir George had no sooner reached the gate of the den, than the lion ran to it, and evinced every demonstration of joy and transport. The animal reared himself up, purred like a cat when pleased, and licked the hand of Sir George, which he had put through the bars. The keeper was astonished and frightened for the safety of his visitor, entreated him not to trust an apparent fit of frenzy, with which the animal seemed to be seized; for he was, without exception, the most fierce and sullen of his tribe which he had ever seen. This, however, had no effect on Sir George, who, notwithstanding every entreaty on the part of the keeper, insisted on entering the lion's den. The moment he got in, the delighted lion threw his paws upon his shoulders, licked his face, and ran about him, rubbing his head on Sir George, purring and fawning like a cat when expressing its affection for its master. This occurrence became the talk of Florence, and reached the ears of the Grand Duke, who sent for Sir George, and requested an interview at the menagerie, that he might witness so extraordinary a circumstance, when Sir George gave the following explanation : 'A captain of a ship from Barbary gave me this lion, when quite a whelp. I brought him up tame; but when I thought him too large to be suffered to run about the house, I built a den for him in my court-yard. From that time he was never permitted to be loose, except when brought to the house to be exhibited to my friends. When he was five years old, he did some mischief by pawing and playing with people in his frolicsome moods. Having griped a man one day a little too hard, I ordered him to be shot, for fear of myself incurring the guilt of what might happen. On this a friend, who happened to be then at dinner with me, begged him as a present. How he came here, I know not.' The Grand Duke of Tuscany, on hearing his story, said it was the very same person who had presented him with the lion."

"Oh! Uncle Thomas: I should have been terribly afraid to have ventured into the lion's den!"

"I dare say you would, John, and so should I. But some stories are recorded of the gentleness of the lion, as almost to justify such acts of what would otherwise appear fool-hardiness.

"Part of a ship's crew being sent ashore on the coast of India for the purpose of cutting wood, the curiosity of one of the men having led him to stray to a considerable distance from his companions, he was much alarmed by the appearance of a large lioness, who made towards him; but, on her coming up, his fear was allayed, by her lying down at his feet, and looking very earnestly, first in his face, and then at a tree some little distance off. After repeating these looks several times, she arose, and proceeded towards the tree, looking back, as if she wished the sailor to follow her. At length he ventured, and, coming to the tree, perceived a huge baboon, with two young cubs in her arms, which he immediately supposed to be those of the lioness', as she crouched down like a cat, and seemed to eye them very stedfastly. The man being afraid to

THE LIONESS AND THE BABOON. Page 117.

ascend the tree, decided on cutting it down, and having his axe with him, he set actively to work, when the lioness seemed most attentive to what he was doing. When the tree fell, she pounced upon the baboon, and, after tearing her in pieces, she turned round, and licked the cubs for some time. She then returned to the sailor, and fawned round him, rubbing her head against him in great fondness, and in token of her gratitude for the service done her. After this, she carried the cubs away one by one, and the sailor rejoined his companions, much pleased with the adventure.

"Another author tells such a graphic story of a lion's entertaining a hunter, that I must let you hear it also, though I must say that I think he has rather overstrained it.

"A hunter on one occasion having gone in search of a lion, and having penetrated a considerable distance into a forest, happened to meet with two whelps of a lion that came to caress him. The hunter stopped with the little animals, and waiting for the coming of the sire or the dam, took out his breakfast, and gave them a part. The lioness arrived, unperceived by the huntsman, so that he had not time, or perhaps wanted the courage, to take his gun. After having for some time looked at the man who was thus feasting her young, the lioness burst away, and soon after returned, bearing with her a sheep, which she came and laid at the huntsman's feet. The hunter, thus become one of the family, took occasion to make a good meal,—skinned the sheep, made a fire, and roasted a part, giving the entrails to the young. The lion, in his turn, came also; and, as if respecting the rights of hospitality, showed no tokens whatever of ferocity. Their guest, the next day, having finished his provisions, returned home, and came to a resolution never more to kill any of these animals, the noble generosity of which he had so fully experienced. He stroked and caressed the whelps at taking leave of them, and the dam and sire accompanied him till he was safely out of the forest."

ESCAPE OF A LIONESS.

"Well, Uncle Thomas, I cannot believe that. I think the man would have been too glad to escape, to have staid so long with such unsafe companions."

"You are quite right, Harry, I cannot expect that you should give credit to a story which I myself disbelieve. Here is a story about the ferocity of the lion, which is, however, beyond all doubt.

"In the year 1816 the horses which were dragging the Exeter mail coach were attacked in the most furious manner by a lioness, which had escaped from a travelling menagerie.

"At the moment when the coachman pulled up, to deliver his bags at one of the stages a few miles from the town of Salisbury, one of the leading horses was suddenly seized by a ferocious animal. This produced a great confusion and alarm. Two passengers, who were inside the mail, got out and ran into the house. The horse kicked and plunged violently; and it was with difficulty the driver could prevent the coach from being overturned. It was soon observed by the coachman and guard, by the light of the lamps, that the animal which had seized the horse was a huge lioness. A large mastiff dog came up and attacked her fiercely, on which she quitted the horse and turned upon him. The dog fled, but was pursued and killed by the lioness, within about forty yards of the place. It appears that the beast had escaped from a caravan, which was standing on the road side, and belonged to a menagerie, on its way to Salisbury fair. An alarm being given, the keepers pursued and hunted the lioness, carrying the dog in her teeth, into a hovel under a granary, which served for keeping agricultural implements. About half past eight, they had secured her effectually, by barricading the place, so as to prevent her escape. The horse, when first attacked, fought with great spirit; and if he had been at liberty, would probably have beaten down his antagonist with his fore feet but in plunging he entangled himself in the harness. The lioness, it appears, attacked him in front, and springing at his throat, had fastened the talons of her fore feet on each side of his gullet, close to the head, while the talons of her hind feet were forced into the chest. In this situation she hung, while the blood was seen streaming, as if a vein had been opened by a lancet. The furious animal missed the throat and jugular vein; but the horse was so dreadfully torn, that he was not at first expected to survive. The expressions of agony, in his tears and moans, were most piteous and affecting. Whether the lioness was afraid of her prey being taken from her, or from some other cause, she continued a considerable time after she had entered the hovel, roaring in a dreadful manner, so loud, indeed, that she was distinctly heard at the distance of half a mile. She was eventually secured and led back in triumph to her cell."

"It was fortunate that it did not attack the passengers, Uncle Thomas."

"Very much so, indeed; it might have turned out a very serious affair, Frank."

CHAPTER VI.

Uncle Thomas tells about the Tiger; its Ferocity and Power, and of the Curious Modes which are adopted for its Capture and Destruction.—Also about the Puma or American Lion, and introduces some Hunting Scenes in North and South America, with other Interesting and Entertaining Adventures.

"Long as the stories were, Boys, which I told you last night about the lion, I have not yet quite done with the animals of the cat kind; there are still one or two stories about the tiger and the puma or American lion, which I wish to tell you of, if you do not think we have already had enough of them."

"Oh, no, Uncle Thomas, pray do continue."

"Very well, I will first tell you about the tiger.

"The tiger, which inhabits the rich jungles of India, nearly equals the lion in strength, and perhaps excels him in activity and ferocity. A very affecting instance of his ferocity, by which a fine young man, the only son of Sir Hector Munro, lost his life, is thus related by one of the party:

"Yesterday morning, Captain George Downey, Lieutenant Pyefinch, poor Mr. Munro (of the Honourable East India Company's service), and myself (Captain Consar), went on shore, on Saugur Island, to shoot deer. We saw innumerable tracks of tigers and deer; but still we were induced to pursue our sport; and did so the whole day. About half past three, we sat down on the edge of the jungle, to eat some cold meat, sent to us from the ship, and had just commenced our meal, when Mr. Pyefinch and a black servant told us there was a fine deer within six yards of us. Captain Downey and I immediately jumped up, to take our guns; mine was nearest, and I had but just laid hold of it, when I heard a roar like thunder, and saw an immense royal tiger spring on the unfortunate Munro, who was sitting down; in a moment his head was in the beast's mouth, and he rushed into the jungle

FEROCITY OF THE TIGER. 125

with him, with as much ease as I could lift a kitten, tearing him through the thickest bushes and trees, every thing yielding to his monstrous strength. The agonies of horror, regret, and, I must say, fear (for there were two tigers), rushed on me at once; the only effort I could make was to fire at him, though the poor youth was still in his mouth. I relied partly on Providence, partly on my own aim, and fired a musket. The tiger staggered, and seemed agitated, which I took notice of to my companions. Captain Downey then fired two shots, and I one more. We retired from the jungle, and, a few minutes after, Mr. Munro came up to us all over blood and fell. We took him on our backs to the boat, and got every medical assistance for him from the Valentine Indiaman, which lay at anchor near the Island, but in vain. He lived twenty-four hours in the utmost torture; his head and skull were all torn and broken to pieces, and he was also wounded, by the animal's claws, all over his neck and shoulders; but it was better to take him away, though irrecoverable, than leave him to be mangled and devoured. We have just read the funeral service over his body, and committed it to the deep. Mr. Munro was an amiable and promising youth. I must observe, there was a large fire blazing close to us, composed of ten or a dozen whole trees. I made it myself on purpose to keep the tigers off, as I had always heard it would. There were eight or ten of the natives about us; many shots had been fired at the place; there was much noise and laughing at the time; but this ferocious animal disregarded all. The human mind cannot form an idea of the scene; it turned my very soul within me. The beast was about four feet and a half high, and nine long. His head appeared as large as that of an ox; his eyes darting fire, and his roar, when he first seized his prey, will never be out of my recollection. We had scarcely pushed our boat from that cursed shore, when the tigress made her appearance, raging, almost mad, and remained on the sand, as long as the distance would allow me to see her."

TIGER HUNTING.

"Oh, dreadful, Uncle Thomas! I declare it makes my hair stand on end!"

"It is a fearful tale, John, and shows you what a scourge such an animal must be to the inhabitants of the country in which it is found. It frequents the deserts of Asia, but in some places where civilization has commenced, it prowls about the villages and commits great havoc among the herds of the inhabitants, who therefore find it necessary to adopt various schemes for its destruction; some of these devices are very curious.

"A large cage of strong bamboos is constructed, and fastened firmly to the ground, in a place which the tigers frequent. In this a man takes up his station for the night. He is generally accompanied by a dog or a goat, which by its extreme agitation is sure to give notice of the tiger's approach. His weapons consist of two or three strong spears, and thus provided he wraps himself in his quilt, and very composedly goes to sleep in the full confidence of safety. By and by the tiger makes his appearance, and after duly reconnoitring all round, begins to rear against the cage, seeking for some means of entering. The hunter, who watches his opportunity, thrusts one of his spears into the animal's body, and seldom fails to destroy it."

"That is a very good plan, Uncle Thomas, and does not seem to be attended with much danger, if the cage be strong enough."

"No, Boys, it is not very dangerous, but I don't think any of you would like to trust yourselves so exposed. Here, however, is another mode of destroying the tiger, which is practised in some parts of India.

"The track of a tiger being ascertained, which though not invariably the same, may yet be known sufficiently for the purpose, the peasants collect a quantity of the leaves of the prous, which are like those of the sycamore, and are common in most underwoods, as they form the largest portion of most jungles in India. These leaves are smeared with a species of bird-lime, made by bruising the berries of a tree by no means scarce. They are then strewed, with the gluten uppermost, near to the spot to which it is understood the tiger usually retires during noon-tide heat. If by chance the animal should tread on one of these smeared leaves his fate is certain. He commences by shaking his paw, with the view to remove the adhesive incumbrance, but finding no relief from that expedient, he rubs the nuisance against his jaw with the same intention, by which means his eyes, ears, &c. become covered with the same substance. This occasions such uneasiness as causes him to roll perhaps among many more smeared leaves, till at length he becomes completely enveloped, and he is deprived of sight, and in this situation may be compared to a man who has been tarred and feathered. The anxiety produced by this strange and novel predicament, soon discovers itself in dreadful howlings, which serve to call the watchful peasants, who in this disabled state find no difficulty in shooting the object of detestation."

"That is better still, Uncle Thomas; I think that is the most ingenious way of catching an animal that I ever heard of."

"I must now tell you something about the puma or American lion, which is also taken in a very ingenious manner by the natives of South America. It is generally hunted by means of dogs. When they unkennel a lion or a tiger, they pursue him till he stops to defend himself. The hunter, who is mounted on a good steed, follows close behind, and if the dogs seize upon the animal, the hunter jumps off his horse, and, while the lion is engaged in contending with the dogs, strikes him on the head, and thus dispatches him. If, however, the dogs are afraid to attack him, the hunter uses his lasso, dexterously fixes it round some part of the animal, and gallops away, dragging it after him. The dogs now rush in and tear him, when he is soon dispatched.

"When wounded the puma grows furious and irresistible. Here is a story which shows the fierceness of the animal:—Two hunters having gone in quest of game to the Catskill mountains, province of New York, each armed with a gun, and accompanied by a dog, they agreed to go in contrary directions round the base of the hill, which formed one of the points of that chain of mountains; and it was settled that, if either discharged his piece, the other should hasten to the spot whence the report proceeded as speedily as possible, to join in the pursuit of whatever game might fall to their lot. They had not been long asunder, when the one heard the other fire, and, agreeably to promise, hastened to join his companion. He looked for him in every direction; but to no purpose. At length, however, he came upon the dog of his friend, dead, and dreadfully lacerated. Convinced by this, that the animal his comrade had shot at was ferocious and formidable, he felt much alarm for his fate, and sought after him with great anxiety. He had not proceeded many yards from the spot where the dog lay prostrate, when his attention was arrested by the ferocious growl of some wild animal. On raising his eyes to the spot whence the sound proceeded, he discovered a large puma couching on the branch of a tree, and under him the body of his friend. The animal's eyes glared at him, and he appeared hesitating whether he should descend, and make an attack on the survivor also, or relinquish his prey, and decamp. The hunter, aware of the celerity of the puma's movements, knew that there was no time for reflection, levelled his piece, and mortally wounded the animal, when it and the body of the man fell together from the tree. His dog then attacked the wounded puma, but a single blow from its paw laid it prostrate. In this state of things, finding his comrade was dead, and knowing it was dangerous to approach the wounded animal, he went in search of assistance, and on returning to the spot he found his companion, the puma, and the two dogs, all lying dead.

"The celebrated naturalist Audubon gives an

PUMA HUNT.

interesting account of a hunt which he had after the puma, in one of the back settlements of North America. In the course of his rambles he arrived at the cabin of a squatter on the banks of Cold-Water River, and after a hospitable reception, and an evening spent in relating their adventures in the chase, it was agreed in the morning to hunt the puma which had of late been making sad ravages among the squatter's pigs.

"The hunters accordingly made their appearance just as the sun was emerging from the horizon. They were five in number, and fully equipped for the chase, being mounted on horses which in some parts of Europe might appear sorry nags, but which, in strength, speed, and bottom, are better fitted for pursuing a puma or bear through woods and morasses than any in that country. A pack of large ugly curs were already engaged in making acquaintance with those of the squatter. He and myself mounted his two best horses, whilst his sons were bestriding others of inferior quality.

"Few words were uttered by the party until we had reached the edge of the swamp where it was agreed that all should disperse and seek for the fresh track of the puma, it being previously settled that the discoverer should blow his horn, and remain on the spot until the rest should join him. In less than an hour, the sound of the horn was clearly heard, and, sticking close the squatter, off we went through the thick woods, guided only by the moon and the repeated call of the distant huntsman. We soon reached the spot, and in a short time the rest of the party came up. The best dog was sent forward to attack the animal, and in a few minutes the whole pack were observed diligently tracking and bearing in their course for the interior of the swamp. The rifles were immediately put in trim, and the party followed the dogs at separate distances, within sight of each other, determined to shoot at no other game than the puma.

"The dogs soon began to mouth, and suddenly quickened their pace. My companions concluded that the beast was on the ground, and putting our horses to a gentle gallop, we followed the curs, guided by their voices. The noise of the dogs increased, when all of a sudden their mode of barking became altered, and the squatter, urging me to push on, told me the beast was *treed*, by which he meant that it had got upon some low branch of a large tree, to rest for a few moments, and that should we not succeed in shooting him while thus situated we might expect a long chase of it. As we approached the spot, we all by degrees united into a body, but on seeing the dogs at the foot of a large tree, separated again, and galloped off to surround it.

" Each hunter now moved with caution, holding his gun ready, and allowing the bridle to dangle on the neck of his horse, as it advanced slowly towards the dogs. A shot from one of th. party was heard, on which the puma was seen to leap to the ground and bound off with such velocity as to show that he was very unwilling to stand our fire longer. The dogs set off in pursuit with the utmost eagerness and a deafening cry; the hunter who had fired came up, and said that his ball had hit the monster, and had probably broken one of his fore legs near the shoulder, the only place at which he could aim; a slight trail of blood was discovered on the ground, but the curs proceeded at such a rate, that we merely noticed this and put spurs to our horses, which galloped on towards the centre of the swamp. One bayou (a part of the swamp in which the water accumulates) was crossed, then another still larger and more muddy, but the dogs were brushing forward, and as the horses began to pant at a furious rate, we judged it expedient to leave them and advance on foot. These determined hunters knew that the animal, being wounded, would shortly ascend another tree, where in all probability he would remain for a considerable time, and that it was easy to follow the track of the dogs. We dismounted, took off the saddles and bridles, set the bells attached to the horses, necks at liberty to jingle, hoppled the animals (fastening the bridle to one of their legs so that they could not stray far), and left them to shift for themselves.

"After marching for a couple of hours, we again heard the dogs. Each of us pressed forward, elated at the thought of terminating the career of the puma; some of the dogs were heard whining, although the greater part barked vehemently. We felt assured that the animal was treed, and that he would rest for some time to recover from his fatigue. As we came up to the dogs we discovered the furious animal lying across a large branch close to the trunk of a cotton-wood tree. His broad breast lay towards us, his eyes were at one time bent on us, and again on the dogs, beneath, and around him; one of his fore-legs hung down loosely by his side, and he lay crouched with his ears lowered close to his head, as if he thought he might remain undiscovered. Three balls were fired at him at a given signal, on which he sprung a few feet from the branch, and tumbled headlong to the ground.

Attacked on all sides by the enraged curs, the infuriated animal fought with desperate valour; but the squatter advancing in front of the party, and almost in the midst of the dogs, shot him immediately behind and beneath the left shoulder. He writhed for a moment in agony, and in another lay dead."

"It must be very exciting employment, hunting the puma, Uncle Thomas."

"And not a little dangerous too, Boys, for you hear how fiercely he maintains his ground. With all their fierceness, however, the fear of man is over even this relentless race of animals. Captain Head, who has written an amusing book called 'Rough Notes of Rapid Rides across the Pampas,' thus speaks on this subject:

"The fear which all wild animals in America have of man is very singularly exhibited in the Pampas. I often rode towards the ostriches and zamas, crouching under the opposite side of my horse's neck; but I always found that, although

A SURPRISE. 139

they would allow my loose horse to approach them, they, even when young, ran from me, though little of my figure was visible ; and when I saw them all enjoying themselves in such full liberty, it was at first not pleasing to observe that one's appearance was every where a signal to them that they should fly from their enemy. Yet it is by this fear 'that man hath dominion over the beasts of the field,' and there is no animal in South America that does not acknowledge this instinctive feeling. As a singular proof of the above, and of the difference between the wild beasts of America and of the old world, I will venture to relate a circumstance which a man sincerely assured me had happened to him in South America.

"He was trying to shoot some wild ducks, and, in order to approach them unperceived, he put the corner of his poncho (which is a sort of long narrow blanket) over his head, and crawling along the ground upon his hand and knees, the poncho not only covered his body, but trailed along the ground behind him. As he was thus creeping by a large bush of reeds, he heard a loud, sudden noise, between a bark and a roar; he felt something heavy strike his feet, and, instantly jumping up, he saw to his astonishment, a large puma actually standing on his poncho; and, perhaps, the animal was equally astonished to find himself in the immediate presence of so athletic a man. The man told me he was unwilling to fire, as his gun was loaded with very small shot; and he therefore remained motionless, the puma standing on his poncho for many seconds; at last the creature turned his head, and walking very slowly away about ten yards, stopped, and turned again: the man still maintained his ground, upon which the puma tacitly acknowledged his supremacy, and walked off."

"I dare say the man was very glad to be so easily quit of such a formidable visitor, Uncle Thomas."

"No doubt of it, Frank. I have one other story to tell you about the puma, which

THE MAN AND THE PUMA. Page 140.

THE PUMA'S GRATITUDE. 141

fortunately exhibits it in a more favourable light than some of those which I have told you.

"During the government of Don Diego de Mendoza, in Paraguay, a dreadful famine raged at Buenos Ayres; yet Diego, afraid to give the Indians a habit of spilling Spanish blood, forbade the inhabitants, on pain of death, to go into the fields, in search of relief, placing soldiers at all the outlets to the country, with orders to fire upon those who should attempt to transgress his orders. A woman, however, called Maldonata, was artful enough to elude the vigilance of the guards, and to effect her escape. After wandering about the country for a long time, she sought shelter in a cavern; but she had scarcely entered it, when she became dreadfully alarmed, on observing a puma occupying the same den. She was, however, soon quieted by the animal approaching and caressing her. The poor brute was very ill, and scarcely able to crawl towards her. Maldonata soon discovered what was the cause of the animal's illness, and kindly ministered to it. It soon recovered, and was all gratitude and attention to its kind benefactress, never returning from searching after its daily subsistence without laying a portion of it at the feet of Maldonata.

"Some time after, Maldonata fell into the hands of the Spaniards; and, being brought back to Buenos Ayres, was conducted before Don Francis Ruez de Galen, who then commanded there. She was charged with having left the city contrary to orders. Galen was a man of a cruel and tyrannical disposition, and condemned the unfortunate woman to a death which none but the most cruel tyrant could have devised. He ordered some soldiers to take her into the country, and leave her tied to a tree, either to perish with hunger, or to be torn to pieces by wild beasts. Two days after, he sent the same soldiers to see what had been her fate, when, to their great surprise, they found her alive and unhurt, though surrounded by pumas and jaguars, while a female puma at her feet kept them at bay. As soon as the puma saw the soldiers, she retired to some distance

THE PUMA'S GRATITUDE. 143

and they unbound Maldonata, who related to them the history of this puma, whom she knew to be the same she had formerly relieved in the cavern. On the soldiers taking Maldonata away, the animal approached, and fawned upon her, as if unwilling to part. The soldiers reported what they had seen to their commander, who could not but pardon a woman who had been so singularly protected, without the danger of appearing more inhuman than pumas themselves.

CHAPTER VII.

Uncle Thomas tells about the Migrating Instinct of Animals.—Of the House Swallow of England; and the Esculent Swallow, whose Nest is eaten by the Chinese.—He tells also about the Passenger Pigeon of America; of the Myriads which are found in various parts of the United States; of the Land-Crab and its Migrations, and of those of the Salmon and the Common Herring.

"UNCLE Thomas, I heard to-day of a swallow which for many years returned to the same window, and built its nest in the same corner. Now as I believe swallows are birds of passage, and leave this country to spend the winter in warmer climates, I wish you to explain to me how it is that they can return from such distances to the same spot."

" That is a question, Frank, which I cannot very well answer, but so many instances of the kind have been observed as to leave no doubt as to the fact. It has sometimes been known even to penetrate into the house, and attach its nest to articles of furniture.

"At Camerton Hall, near Bath, a pair of swallows built their nest on the upper part of the frame of an old picture over the chimney; and, coming into the room through a broken pane in one of the windows, they continued to use the same place for their nest for three successive years, and would probably have continued to do so, but the room having been put into repair, they could no longer obtain access to it."

"Is it want of food which makes birds migrate, Uncle Thomas?"

"Principally, I should say that it is so, Frank, but in shifting from one place to another they only fulfil an instinct impressed on them by their Creator for the preservation of their species. Thus, for instance, an old swallow might know by experience, that when its food fails here, it begins to become plentiful elsewhere, but the young bird which had never been more than a few miles from the place where it was hatched, can have no such experimental knowledge; yet, when the season arrives, we find the whole flock ready to set out. I dare say you have all seen them, Boys, gathering in flocks and resting on the house tops, as if taking breath before setting out on their long journey."

"Oh, yes, Uncle Thomas, I have often seen them doing so, but I have heard that they dive to the bottom of lakes and ponds, and remain there till winter is over."

"Many foolish stories are told of swallows being found in such situations, Harry, but they are now well known to be fables. There is no doubt that they migrate in the same way as many other birds. Last autumn I watched with great pleasure the movements of a flock, which was evidently preparing for their arduous flight.

"For several evenings they assembled in large numbers on a tree at a short distance from my house, and, after remaining seated for some time, one of them, who appeared to be commander-in-chief, kept flying about in all directions, and at length, with a sharp and loudly repeated call, he darted up into the air. In an instant the whole congregation were on the wing, following their leader in a sort of spiral track. In a little time they had risen so high that I lost sight of them, but after a short absence they again returned and took up their position on the tree which they had just left.

"This manœuvre they continued for some time, till one day they set off in reality, and I saw no more of them for the winter."

"I read, somewhere, Uncle Thomas, that the Chinese eat swallows' nests. I cannot understand this, Sir; surely the mud and clay, of which swallows' nests are composed, would make but an indifferent repast."

"I dare say they would, Frank, if they were made of clay and mud, as the nests of our swallows are; but such is not the case. Various opinions are entertained as to the substance of which the nest of the esculent swallow is made. Sir George Staunton, who accompanied Lord Macartney in his embassy to China, gives a very interesting account both of the swallow and of its nest.

"In the Cass," says Sir George, "a small island near Sumatra, we found two caverns running horizontally into the side of the rock, and in these were a number of those birds' nests so much prized by the Chinese epicures. They seemed to be composed of fine filaments, cemented together by a transparent viscous matter, not unlike what is left by the foam of the sea upon stones alternately covered by the tide, or those gelatinous animal substances found floating on every coast. The nests adhere to each other and to the sides of the cavern, mostly in horizontal rows, without any break or interruption, and at different depths from fifty to five hundred feet. The birds that build these nests are small grey swallows, with bellies of a dirty white. They were flying about in considerable numbers, but were so small, and their flight was so quick, that they escaped the shot fired at them. The same sort of nests are said to be also found in deep caverns at the foot of the highest mountains in the middle of Java, at a distance from the sea; from which source it is thought that the birds derive no materials, either for their food, or the construction of their nests, as it does not appear probable they should fly in search of either over the intermediate mountains, which are very high, or against the boisterous winds prevailing thereabouts. They feed on insects, which they find hovering over stagnated pools between the mountains, and for the catching of which their wide opening beaks are particularly adapted. They prepare their nests from the best remnants of their food. Their greatest enemy is the kite, who often intercepts them in their passage to and from the caverns, which are generally surrounded with rocks of grey limestone or white marble. The colour and value of the nests depend on the quantity and quality of the insects caught, and perhaps also on the situation where they are built. Their value is chiefly ascertained by the uniform fineness and delicacy of their texture; those that are white and transparent being most esteemed, and fetching often in China their weight in silver.

"These nests are a considerable object of traffic among the Javanese, many of whom are employed in it from their infancy. The birds, after having spent nearly two months in preparing their nests, lay each two eggs, which are hatched in about fifteen days. When the young birds become fledged, it is thought the proper time to seize upon their nests, which is done regularly three times a year, and is effected by means of ladders of bamboo and reeds, by which the people descend into the caverns; but when these are very deep, rope-ladders are preferred. This operation is attended with much danger, and several perish in the attempt. The inhabitants of the mountains generally employed in this business begin always by sacrificing a buffalo, which custom is observed by the Javanese

THE ESCULENT SWALLOW. 151

on the eve of every extraordinary enterprise. They also pronounce some prayers, anoint themselves with sweet-scented oil, and smoke the entrance of the cavern with gumbenjamin. Near some of the caverns a tutelar goddess is worshipped, whose priest burns incense, and lays his projecting hands on every person preparing to descend. A flambeau is carefully prepared at the same time, with a gum which exudes from a tree growing in the vicinity, and which is not easily extinguished by fixed air or subterraneous vapours."

"But how are the nests eaten, Uncle Thomas? Are they prepared in any way, or are they fit for use as they are taken down?"

"They are always prepared before they are eaten. The finest sort, which are of a clear colour, and not unlike isinglass, are dissolved in broth, to which they are said to give an exquisite flavour. After being soaked, they are sometimes introduced into the body of a fowl and stewed; but I am not quite versed in all the mysteries of a Chinese kitchen, so you must be satisfied with these two modes of preparation."

"Thank you, Uncle Thomas."

"I have only one more story to tell you about the swallow, Boys, and then I must turn to two or three other animals, whose peregrinations exhibit as strong instances of instinct as it does."

"A swallow's nest, built in the west corner of a window facing the north, was so much softened by the rain beating against it, that it was rendered unfit to support the superincumbent load of five pretty full grown swallows. During a storm the nest fell into the lower corner of the window, leaving the young brood exposed to all the fury of the blast. To save the little creatures from an untimely death, the owner of the house benevolently caused a covering to be thrown over them, till the severity of the storm was past. No sooner had it subsided, than the sages of the colony assembled, fluttering round the window, and hovering over the temporary covering of the fallen nest. As soon as this careful anxiety was observed, the covering was removed, and the utmost joy evinced by the group, on finding the young ones alive and unhurt. After feeding them, the members of this assembled community arranged themselves into working order. Each division taking its appropriate station, commenced instantly to work, and before night-fall they had jointly completed an arched canopy over the young brood in the corner where they lay, and securely covered them against a succeeding blast. Calculating the time occupied by them in performing this piece of architecture, it appeared evident that the young must have perished from cold and hunger, before any single pair could have executed half the job."

"How very kind, Uncle Thomas! Had they been reasoning creatures, they could not have behaved more properly."

"I dare say not, Frank. Such traits overstep the limits of *instinct*, and almost trespass on that of reason."

"You asked, Frank, if it was want of food which prompted the flight of migratory animals from one place to another. In some cases it is so, undoubtedly; as for instance, in that which I am now going to tell you about, the American passenger pigeon; it is from the work of the great naturalist, Wilson.

"The migrations of these pigeons appear to be undertaken rather in quest of food than merely to avoid the cold of the climate; since we find them lingering in the northern regions around Hudson's Bay so late as December, and since their appearance is so casual and irregular, sometimes not visiting certain districts for several years in any considerable numbers, while at other times they are innumerable. I have often witnessed these migrations in the Genesee country, often in Pennsylvania, and also in various parts of Virginia, with amazement; but all that I have seen of them are mere straggling parties, when compared with the congregated millions which I have since beheld in the western forests in the states of Ohio, Kentucky, and the Indiana territory. These fertile and extensive regions abound with the nutritious beech nut, which constitutes the chief food of the wild pigeon. In seasons when these nuts are abundant, corresponding multitudes of pigeons may be confidently expected. It sometimes happens, that having consumed the whole produce of the beech trees in an extensive district, they discover another at the distance of perhaps sixty or eighty miles, to which they regularly repair every morning, and return as regularly in the course of the day, or in the evening, to their place of general rendezvous, or, as it is generally called, the roosting place. These roosting places are always in the wood, and sometimes occupy a large extent of forest. When they have frequented one of these places for some time, the appearance it exhibits is surprising. The ground is covered to the depth of several inches with their droppings ; all the tender grass and underwood destroyed ; the surface strewed with large limbs of trees, broken down by the weight of the birds clustering one above another, and the trees themselves, for thousand of acres, killed as completely as if girdled with an axe. The marks of this desolation remain for many years on the spot, and numerous places could be pointed out, where for several years after scarcely a single vegetable made its appearance.

"When these roosts are first discovered, the inhabitants from considerable distances visit them in the night with guns, clubs, long poles, pots of sulphur, and various other engines of destruction. In a few hours they fill many sacks, and load their horses with them. By the Indians, a pigeon roost or breeding place is considered an important source of national profit and dependence for the season, and all their active ingenuity is exercised on the occasion. The breeding place differs from the former in its greater extent. In the western countries before mentioned, these are generally in beech woods, and often extend in nearly a straight line across the country for a great way. Not far from Shelbyville, in the state of Kentucky, about five years ago, there was one of these breeding places, which stretched through the woods in nearly a north and south direction, which was several miles in breadth, and was said to be upwards of forty miles in extent. In this tract almost every tree was furnished with nests wherever the branches could accommodate them. The pigeons made their first appearance there about the 10th of April, and left it altogether with their young before the 25th of May.

"As soon as the young were fully grown, and before they left the nests, numerous parties of the inhabitants, from all parts of the adjacent country, came with waggons, axes, beds, cooking utensils, many of them accompanied by the greater part of their families, and encamped for several days in this immense nursery. Several of them informed me that the noise in the woods was so great as to terrify their horses, and that it was difficult for one person to hear another speak without bawling in his ear. The ground was strewed with broken limbs of trees, eggs, and young squab pigeons which had been precipitated from above, and on which herds of hogs were fattening; hawks, buzzards, and eagles were sailing about in great numbers, and seizing the squabs from their nests at pleasure; while from twenty feet upwards to the tops of the trees, the view through the woods presented a perpetual tumult of crowding and fluttering multitudes of pigeons, their wings roaring like thunder, mingled with the frequent crash of falling timber; for now the axemen were at work, cutting down those trees which seemed to be most crowded with nests, and contrived to fell them in such a manner, that in their descent they might bring down several others, by which means the falling of one large tree sometimes produced two hundred squabs, little inferior in size to the old pigeons, and almost one mass of fat. On some single trees, upwards of one hundred nests were found, each containing one young only, a circumstance in the history of this bird not generally known to naturalists. It was dangerous to walk under these fluttering and flying millions, from the frequent fall of large branches, broken down by the weight of the multitudes above, and which in their descent often destroyed numbers of the birds themselves.

"I had left the public road to visit the remains of the breeding place near Shelbyville, and was traversing the woods with my gun on my way to Frankfort, when about one o'clock, the pigeons which I had observed flying the greater part of the morning northerly, began to return in such immense numbers as I never before had witnessed; coming to an opening by the side of a creek called the Benson, where I had a more uninterrupted view, I was astonished at their appearance. They were flying with great steadiness and rapidity, at a height beyond gun-shot, and several strata deep, and so close together, that could shot have reached them, one discharge could not have failed in bringing down several individuals. From right to left as far as the eye could reach, the breadth of this vast procession extended, seeming every where equally crowded. Curious to determine how long this appearance would continue, I took out my watch to note time, and sat down to observe them. It was then half-past one; I sat for more than an hour, but instead of a diminution of this prodigious procession, it seemed rather to increase both in numbers and rapidity, and anxious to reach Frankfort before night, I arose and went on. About four o'clock in the afternoon, I crossed the Kentucky river at the town of Frankfort, at which time the living torrent above my head seemed as numerous and as extensive as ever; and long after this, I observed them in large bodies that continued to pass for six or eight minutes, and these again were followed by other detached bodies, all moving in the same south-east direction, till after six in the evening. The great breadth of front which this mighty multitude preserved would seem to intimate a corresponding breadth of their breeding place, which, by several gentlemen who had lately passed through part of it, was stated to me at several miles. It was said to be in Green County, and that the young began to fly about the middle of March. On the 17th of April, forty-nine miles beyond Danville, and not far from Green River, I crossed this same breeding place, where the nests for more than three miles spotted every tree; the leaves not being yet out, I had a fair prospect of them, and was really astonished at their numbers. A few bodies of pigeons lingered yet in different parts of the woods, the roaring of whose wings was heard in various quarters around me.

"The vast quantity of food which these multitudes consume is a serious loss to the other animals, such as bears, pigs, squirrels, which are dependent on the fruits of the forest. I have taken from the crop of a single wild pigeon a good handful of the kernels of beech nuts intermixed with acorns and chesnuts. To form a rough estimate of the daily consumption of one of these immense flocks, let us first attempt to calculate the numbers above mentioned, as seen in passing between Frankfort and the Indian Territory. If we suppose this column to have been one mile in breadth (and I believe it to have been much more), and that it moved at the rate of one mile in a minute, four hours, the time it continued passing, would make its whole length two hundred and forty miles. Again, supposing that each square yard of this moving body comprehended three pigeons, the square yards in the whole space, multiplied by three, would give two thousand two hundred and thirty millions two hundred and seventy-two thousand pigeons!—an almost incredible multitude, and yet far below the actual amount. Computing each of these to consume half a pint of mast (nuts, and other seeds of trees) daily, the whole quantity, at this rate, would equal seventeen millions four hundred and twenty-four thousand bushels per day! Heaven has wisely and graciously given to these birds rapidity of flight, and a disposition to range over vast uncultivated tracts of the earth; otherwise they must have perished in the districts where they resided, or devoured the whole productions of agriculture, as well as those of the forests.

"The appearance of large detached flocks of these birds in the air, and the various evolutions they display, are strikingly picturesque and interesting. In descending the Ohio by myself, I often rested on my oars to contemplate their aerial manœuvres. A column of eight or ten miles in length would appear from Kentucky high in air, steering across to Indiana. The leaders of this great body would sometimes gradually vary their course, till it formed a large bend of more than a mile in diameter, those behind tracing the exact route of their predecessors. This would continue sometimes long after both extremities were beyond the reach of sight; so that the whole with its glittering undulations marked a space on the face of the heavens resembling the windings of a vast majestic river. When this bend became very great, the birds, as if sensible of the unnecessarily circuitous route they were taking, suddenly changed their direction, so that what was in column before became an immense front, straightening all its indentures until it swept the heavens in one vast and infinitely extended line. Other lesser bodies also united with each other as they happened to approach, and with such ease and elegance of evolution, forming new figures and varying these as they united or separated, that I was never tired of contemplating them. Sometimes a hawk would sweep on a particular part of the column from a great height, when almost as quick as lightning that part shot downwards out of the common track, but soon rising again, continued advancing at the same height as before. This inflection was continued by those behind, who, on arriving at this point, dived down almost perpendicularly to a great depth, and, rising, followed the exact path of those that went before.

"Happening to go ashore one charming afternoon to purchase some milk at a house that stood near the river, and while talking with the people within doors, I was suddenly struck with astonishment at a loud rushing roar, succeeded by instant darkness, which for the first moment I took for a tornado about to overwhelm the house, and every thing around, in destruction. The people observing my surprise, coolly said, 'It is only the pigeons,' and on running out, I beheld a flock thirty or forty yards in width, sweeping along very low between the house and the mountain or height that formed the second bank of the river. These continued crossing for more than a quarter of an hour, and at length varied their bearing, so as to pass over the mountain, behind which they disappeared before the rear came up."

"That is amazing, Uncle Thomas; two thousand millions of live birds! I can scarcely form an adequate idea of such a mass of living creatures."

"There is something almost overwhelming in

the idea, Frank; and yet in some parts of the world are to be found flocks of animals hardly less surprisingly numerous, when we consider how much less they are fitted for moving about, travelling at stated intervals from the mountains to the sea coast, and returning again to their old habitations, after having fulfilled the purposes for which this instinctive feeling was implanted in them."

"Which animals do you mean, Uncle Thomas?"

"I allude to the land-crab, which is a native of the Bahamas, and also of most of the other islands between the tropics. They live in clefts of the rocks, or holes which they dig for themselves among the mountains, and subsist on vegetables. About the months of April and May, they descend to the sea coast in a body of millions at a time, for the purpose of depositing their spawn. They march in a direct line towards their destination, and seldom turn out of their way, even should they encounter a wall or a house, but boldly attempt to scale it. If, however, they arrive at a river, they wind along the course of the stream, and thus reach the sea.

"In their procession they are as regular as an army under the command of an experienced general, and are usually divided into three battalions. The first body consists of the strongest males, which march forward to clear the route and face the greatest dangers. The main body is composed of females, which are formed into columns, sometimes extending fifty or sixty yards in breadth and three miles in depth. Three or four days after these follows the third division or rear guard, a straggling undisciplined tribe, consisting both of males and females, but neither so robust nor so vigorous as the former.

"Though easily drowned, a certain proportion of moisture seems necessary to the existence of these animals, and the advanced guard is often obliged to halt from the want of rain. The females, indeed, never leave the mountains till the rainy season has fairly set in. They march chiefly during the night, but if it happens to rain during the day, they always profit by it. When the sun is hot they halt till evening. They march very slowly, and are sometimes three months in gaining the shore. When alarmed they run in a confused and disorderly manner, holding up and clattering their nippers with a threatening attitude, and if suffered to take hold of the hand they bite severely. If in their journey any of them should be so maimed as to be unable to proceed, the others fall upon it and devour it.

"Arrived at the coast, they prepare to cast their spawn. They go to the edge of the water, and suffer the waves to wash twice or thrice over their bodies, and then withdraw to seek a lodging upon the land. After a short time the spawn becomes ready for being deposited, when they again seek the sea-side, and leave the spawn to be brought to maturity by the heat of the sun. Much of the spawn, which exactly resembles the roe of a herring, is devoured by the fishes; that which escapes soon arrives at maturity, and millions of little crabs are then to be seen slowly travelling towards the mountains.

"The old ones in the mean time seek to return to their old haunts, but so feeble are they that they seem scarcely able to crawl along. Some of them, indeed, are obliged to remain in the level parts of the country till they recover, making holes in the earth, which they block up with leaves and dirt. In these they cast their old shells, after which they soon recover, and become so fat as to be delicious food.

"At the season of their descent from the mountains, the natives of the islands which they inhabit, eagerly wait for them and destroy them in thousands. On their descent they are only taken for the roe or spawn, the flesh being then poor and lean : on their return from the sea-side they are in greatest repute, being then fat and high flavoured.

"The crab-catchers adopt various modes of securing them, but they are obliged to be very cautious, for when the animals perceive themselves attacked, they throw themselves on their back, and snap their claws about, pinching whatever they lay hold of very severely. The crab-catchers, however, soon learn to seize them by the hind legs, in such a manner as that the nippers cannot reach them."

" You said, Uncle Thomas, that the fishes watched the descent of the crabs, that they might feed on the spawn. Do you think that they are endowed with reasoning powers, as well as the higher classes of animals, Uncle Thomas?"

" No doubt of it, Frank. Old Isaac Walton, the most amusing author on angling who ever wrote, tells many curious stories about fishes, of their coming to be fed at the sound of a bell, and so forth.

" Many fishes exhibit the migratory instinct quite as distinctly as those animals which I have just told you about. The salmon leaves the sea, and seeks its way up the rivers, stemming their most rapid currents, and scaling highest waterfalls with a pertinacity which can only be the result of an instinct implanted in them by their Creator."

"And the herring, Uncle Thomas; does not it come every year from the Polar seas to spawn on our shores? I read a very interesting account of their progress southwards somewhere lately."

"I can tell you where, Frank; I will show it you, and when you have read it aloud, I will point out one or two mistakes, which it is as well to clear your mind of. It is in old Pennant's work; here it is; will you read it to us, John?"

"With pleasure, Uncle Thomas.

"This mighty army begins to put itself in motion in the spring. They begin to appear off the Shetland Islands in April and May. This is the first check this army meets in its march southward. There it is divided into two parts; one wing of those destined to visit the Scottish coast takes to the east, the other to the western shores of Great Britain, and fill every bay and creek with their numbers; others proceed towards Yarmouth, the great and ancient mart of herrings; they then pass through the British channel, and after that in a manner disappear. Those which take to the west, after offering themselves to the Hebrides, where the great stationary fishery is, proceed towards the north of Ireland, where they meet with a second interruption and are obliged to make a second division; the one takes to the western side and is scarcely perceived, being soon lost in the immensity of the Atlantic, but the other, which passes into the Irish sea, rejoins, and feeds the inhabitants of most of the coasts that border on it. The brigades, as we call them, which are separated from the greater columns, are often capricious in their motions, and do not show an invariable attachment to their haunts."

"Thank you, John. Now all this sounds very fine, and seems very systematic. It has but one objection—it is quite untrue. It is in the first place more than doubtful if the herring frequents the Polar seas at all; and in the second place, the most distinguished naturalists are of opinion that it never leaves the neighbourhood of our own shores, but merely retires to the deep water after it has spawned, and there remains till the return of another season calls it again to the shores to undergo a similar operation. So you see, Frank, it does not follow that an interesting account of an animal's habits is necessarily a true one."

CHAPTER VIII.

Uncle Thomas tells about the Baboons, and their Plundering Excursions to the Gardens at the Cape of Good Hope, also about Le Vaillant's Baboon, Kees, and his Peculiarities; the American Monkeys; and relates an amusing Story about a young Monkey deprived of its Mother, putting itself under the Fostering Care of a Wig-Block.

"Oh, Uncle Thomas, I saw such a strange looking creature to-day. It was so ugly. It seemed to be a very large monkey, it was as big as a boy."

"I heard of it, Boys, though I did not see it. It was a baboon, and one of the largest of the species.—It was what is called the dog-faced baboon."

"Where do such animals come from, Uncle Thomas."

"From Africa, John, and I believe they are not to be found elsewhere. They are very fierce and mischievous creatures, and are said sometimes even to attack man, but this I believe to be an exaggeration. Immense troops of them inhabit the mountains in the neighbourhood of the Cape of Good Hope, whence they descend in bands to plunder the gardens and orchards. In these excursions they move on a concerted plan, placing sentinels on commanding spots to give notice of the approach of an enemy. On the appearance of danger, the sentinel utters a loud yell, upon which the whole troop retreats with the utmost precipitation."

"Do they carry the spoil with them when they are thus disturbed, Uncle Thomas?"

"When disturbed they are said to break in pieces the fruit which they have gathered, and cram it into their cheek pouches—receptacles with which nature has furnished them for keeping articles of food till they are wanted.

"Le Vaillant, a traveller in Africa, had a dog-faced baboon which accompanied him on his journey, and he found its instinct of great service to him in various ways. As a sentinel he was better than any of the dogs. So quick was his sense of danger, that he often gave notice of the approach of beasts of prey, when every thing else seemed sunk in security. He was also very useful in guarding the people of the expedition from danger, from using unwholesome or poisonous fruits. The animal's name was Kees. Here is the very interesting account which his master gives of him.

"Whenever we found fruits or roots, with which my Hottentots were unacquainted, we did not touch them till Kees had tasted them. If he threw them away, we concluded that they were either of a disagreeable flavour, or of a pernicious quality, and left them untasted. The ape possesses a peculiar property, wherein he differs greatly from other animals, and resembles man,—namely, that he is by nature equally gluttonous and inquisitive. Without necessity, and without appetite, he tastes every thing that falls in his way, or that is given to him. But Kees had a still more valuable quality,—he was an excellent sentinel; for, whether by day or night, he immediately sprang up on the slightest appearance of danger. By his cry, and the symptoms of fear which he exhibited, we were always apprized of the approach of an enemy, even though the dogs perceived nothing of it. The latter at length learned to rely upon him with such confidence, that they slept on in perfect tranquillity. I often took Kees with me when I went a hunting; and when he saw me preparing for sport, he exhibited the most lively demonstrations of joy. On the way he would climb into the trees, to look for gum, of which he was very fond. Sometimes he discovered to me honey. deposited in the clefts of rocks, or hollow trees. But if he happened to have met with neither honey nor gum, and his appetite had become sharp by his running about, I always witnessed a very ludicrous scene. In those cases, he looked for roots, which he ate with great greediness, especially a particular kind, which, to his cost, I also found to be very well tasted and refreshing, and therefore insisted upon sharing with him. But Kees was no fool. As soon as he found such a root, and I was not near enough to seize upon my share of it, he devoured it in the greatest haste, keeping his eyes all the while riveted on me. He accurately measured the distance I had to pass before I could get to him; and I was sure of coming too late. Sometimes, however, when he had made a mistake in his calculation, and I came upon him sooner than he expected, he endeavoured to hide the root, in which case I compelled him, by a box on the ear, to give me up my share. But this treatment caused no malice between us; we remained as good friends as ever. In order to draw these roots out of the ground, he employed a very ingenious method, which afforded me much amusement. He laid hold of the herbage with his teeth, stemmed his fore feet against the ground, and drew back his head, which gradually pulled out the root. But if this expedient, for which he employed his whole strength, did not succeed, he laid hold of the leaves as before, as close

THE BABOON. 179

to the ground as possible, and then threw himself heels over head, which gave such a concussion to the root, that it never failed to come out.

"When Kees happened to tire on the road, he mounted upon the back of one of my dogs, who was so obliging as to carry him whole hours. One of them, which was larger and stronger than the rest, hit upon a very ingenious artifice, to avoid being pressed into this piece of service. As soon as Kees leaped upon his back he stood still, and let the train pass, without moving from the spot. Kees still persisted in his intention, till we were almost out of his sight, when he found himself at length compelled to dismount, upon which both the baboon and dog exerted all their speed to overtake us. The latter, however, gave him the start, and kept a good look-out after him, that he might not serve him in the same manner again. In fact, Kees enjoyed a certain authority with all my dogs, for which he perhaps was indebted to the superiority of his instinct. He could not endure a competitor;

if any of the dogs came too near him when he was eating, he gave them a box on the ear, which compelled him immediately to retire to a respectful distance.

"Serpents excepted, there were no animals of whom Kees stood in such great dread as of his own species,—perhaps owing to a consciousness, that he had lost a portion of his natural capacities. Sometimes he heard the cry of the other apes among the mountains, and, terrified as he was, he yet answered them. But if they approached nearer, and he saw any of them, he fled, with a hideous cry, crept between our legs, and trembled over his whole body. It was very difficult to compose him, and it required some time before he recovered from his fright.

"Like all other domestic animals, Kees was addicted to stealing. He understood admirably well how to loose the strings of a basket, in order to take victuals out of it, especially milk, of which he was very fond. My people chastised him for these thefts; but that did not make him amend his conduct. I myself sometimes whipped him; but then he ran away, and did not return again to the tent, until it grew dark. Once as I was about to dine, and had put the beans which I had boiled for myself upon a plate, I heard the voice of a bird, with which I was not acquainted. I left my dinner standing, seized my gun, and run out of my tent. After the space of about a quarter of an hour, I returned, with the bird in my hand; but to my astonishment, found not a single bean upon the plate. Kees had stolen them all, and taken himself out of the way. When he had committed any trespass of this kind, he used always, about the time when I drank tea, to return quietly, and seat himself in his usual place, with every appearance of innocence, as if nothing had happened; but this evening he did not let himself be seen; and on the following day, also, he was not seen by any of us; and in consequence, I began to grow seriously uneasy about him, and apprehensive that he might be lost for ever, but on the third day, one of my people, who had been to fetch water, informed me that he had seen Kees in the neighbourhood; but that as soon as the animal espied him, he had concealed himself again. I immediately went out and beat the whole neighbourhood with my dogs. All at once, I heard a cry, like that which Kees used to make when I returned from my shooting, and had not taken him with me. I looked about, and at length espied him, endeavouring to hide himself behind the large branches of a tree. I now called to him in a friendly tone of voice, and made motions to him to come down to me. But he would not trust me, and I was obliged to climb up the tree to fetch him. He did not attempt to fly, and we returned together to my quarters; here he expected to receive his punishment; but I did nothing, as it would have been of no use.

"When exhausted with the heat of the sun, and the fatigues of the day, with my throat and mouth covered with dust and perspiration, I was ready to sink gasping to the ground, in tracts destitute of shade, and longed even for the dirtiest ditch-water; but after seeking long in vain, lost all hopes of finding any in the parched soil. In such distressing moments, my faithful Kees never moved from my side. We sometimes got out of our carriage, and then his sure instinct led him to a plant. Frequently the stalk was fallen off, and then all his endeavours to pull it out were in vain. In such cases, he began to scratch in the earth with his paws; but as that would also have proved ineffectual, I came to his assistance with my dagger, or my knife, and we honestly divided the refreshing root with each other.

"An officer, wishing one day to put the fidelity of my baboon, Kees, to the test, pretended to strike me. At this Kees flew in a violent rage, and, from that time, he could never endure the sight of the officer. If he only saw him at a distance, he began to cry and make all kinds of grimaces, which evidently showed that he wished to revenge the insult that had been done to me; he ground his teeth, and endeavoured, with all his might, to fly at his face, but that was out of his power, as he was chained down. The offender several times endeavoured, in vain, to conciliate him, by offering him dainties, but he remained long implacable.

"When any eatables had been pilfered at my quarters, the fault was always laid first upon Kees; and rarely was the accusation unfounded. For a time the eggs which a hen laid me were constantly stolen away, and I wished to ascertain whether I had to attribute this loss also to him. For this purpose, I went one morning to watch him, and waited till the hen announced by her cackling that she had laid an egg. Kees was sitting upon my vehicle; but the moment he heard the hen's voice he leapt down, and was running to fetch the egg. When he saw me he suddenly stopped, and affected a careless posture, swaying himself backwards upon his hind legs, and assuming a very innocent look; in short, he employed all his art to deceive me with

THE DOG AND BABOON. Page 185.

respect to his design. His hypocritical manœuvres only confirmed my suspicions, and, in order in my turn to deceive him, I pretended not to attend to him, and turned my back to the bush where the hen was cackling, upon which he immediately sprang to the place. I ran after him, and came up to him at the moment when he had broken the egg, and was swallowing it. Having caught the thief in the fact, I gave him a good beating upon the spot; but this severe chastisement did not prevent his soon stealing fresh-laid eggs again. As I was convinced that I should never be able to break Kees of his natural vices, and that, unless I chained him up every morning, I should never get an egg, I endeavoured to accomplish my purpose in another manner: I trained one of my dogs, as soon as the hen cackled, to run to the nest, and bring me the egg without breaking it. In a few days the dog had learned his lesson; but Kees, as soon as he heard the hen cackle, ran with him to the nest. A contest now took place between them, who should have the egg; often the dog was foiled, although he was the stronger of the two. If he gained the victory, he ran joyfully to me with the egg, and put it into my hand. Kees, nevertheless, followed him, and did not cease to grumble and make threatening grimaces at him, till he saw me take the egg,—as if he was comforted for the loss of his booty by his adversary's not retaining it for himself. If Kees had got hold of the egg, he endeavoured to run with it to a tree, where, having devoured it, he threw down the shells upon his adversary, as if to make game of him. In that case, the dog returned, looking ashamed, from which I could conjecture the unlucky adventure he had met with.

"Kees was always the first awake in the morning, and when it was the proper time he awoke the dogs, who were accustomed to his voice, and, in general, obeyed without hesitation the slightest motions by which he communicated his orders to them, immediately taking their posts about the tent and carriage, as he directed them."

THE BABOON.

"What a delightful companion Kees must have been, Uncle Thomas!"

"He must at least have been an amusing one, Frank, and not an unuseful one either. There are, however, great variations in this respect among the monkeys; some of them are most lively creatures, seldom sitting still for a couple of minutes, while others are retired and gloomy in their dispositions, and some are most fickle and uncertain. The fair monkey, though one of the most beautiful of the tribe, is of the latter description, as the following story will testify:—

"An animal of this class, which from its extreme beauty and gentleness was allowed to ramble at liberty about a ship, soon became a great favourite among the crew, and in order to make him perfectly happy, as they imagined, they procured him a wife. For some weeks he was a devoted husband, and showed her every attention and respect. He then grew cool, and became jealous of any kind of civility shown her by the master of the vessel, and began to use her with much cruelty. His treatment made her wretched and dull; and she bore the spleen of her husband with that fortitude which is characteristic of the female sex of the human species. And pug, like the lords of the creation, was up to deceit, and practised pretended kindness to his spouse, to effect a diabolical scheme, which he seemed to premeditate. One morning, when the sea ran very high, he seduced her aloft, and drew her attention to an object at some distance from the yard-arm; her attention being fixed, he all of a sudden applied his paw to her rear, and canted her into the sea, where she fell a victim to his cruelty. This seemed to afford him high gratification, for he descended in great spirits."

"Oh, what a wretched creature, Uncle Thomas. I wonder the sailors did not throw him into the sea also."

"Stay, Frank, you are somewhat too hasty. He deserved certainly to be punished; but I doubt whether it would have been proper to have put him to death for his misdeed. All monkeys are not, however, equally cruel; some of them, indeed, are remarkable for the instinctive kindness which they evince towards their young. When threatened by danger, they mount them on their back, or clasp them firmly to their breast, to which the young creatures secure themselves, by means of their long and powerful arms, so as to permit of their parent moving about, and springing from branch to branch, with nearly as much facility as if she were perfectly free from all incumbrance."

" Oh, I can readily believe that, Uncle Thomas. One day lately, at the Zoological Gardens, I saw two monkeys clasping a young one between them, to keep it warm. They seemed so fond of it."

" Yes, Frank, I have also seen them occupied in the same way. I was quite delighted at such an unexpected exhibition of tenderness.

" Some of the monkeys which are natives of the American continent have the singular characteristic of prehensile tales; that is, of tails which they can more about, and lay hold of branches of trees with nearly as much ease as they can with their hands. The facilities which this affords them for moving about with celerity among the branches of trees is astonishing. The firmness of the grasp which it takes of the tree is no less surprising, for if it makes a single coil round a branch, it is quite sufficient, not only to support the weight of the animal, but to enable it to swing in such a manner as to gain a fresh hold with its feet."

"That is very curious, Uncle Thomas. Is there any other animal which has this power in the tail."

"Oh, yes, Frank, several of the lizards have the power, as well as some other animals; the little harvest mouse, for instance; but none of them are possessed of it in so high a degree as the American monkeys.

"I have now pretty well exhausted my stories about the monkey tribe. I recollect only one more at present, and it occurred to the same traveller to whom Kees belonged.

" In one of his excursions he happened to kill a female monkey, which carried a young one on her back. The little creature, as if insensible of its mother's death, continued to cling to the dead body till they reached their evening quarters; and even then it required considerable force to disengage it. No sooner, however, did the little creature feel itself alone, than it darted towards a wooden block, on which was placed the wig of Le Vaillant's father, mistaking it for its dead mother. To this it clung most pertinaciously by its fore paws; and such was the force of this deceptive instinct, that it remained in the same position for about three weeks, all this time evidently mistaking the wig for its mother. It was fed, from time to time, with goat's milk; and, at length, emancipated itself voluntarily, by quitting the fostering care of the peruke. The confidence which it ere long assumed, and the amusing familiarity of its manners, soon rendered it a favourite. The unsuspecting naturalist had, however, introduced a wolf in sheep's clothing into his dwelling: for. one morning, on entering his chamber the door of which had been imprudently left open, he beheld his young favourite making a hearty breakfast on a very noble collection of insects. In the first transports of his anger, he resolved to strangle the monkey in his arms: but his rage immediately gave way to pity, when he perceived that the crime of its voracity had carried the punishment along with it. In eating the beetles, it had swallowed several of the pins on which they were transfixed. Its agony, consequently, became great; and all his efforts were unable to preserve its life."

"Poor creature! How unfortunate, Uncle Thomas. It must, however, have been a very stupid animal to mistake a wig for its mother."

CHAPTER IX.

Uncle Thomas concludes Stories about Instinct with several Interesting Illustrations of the Affections of Animals, particularly of the Instinct of Maternal Affection, in the course of which he narrates the Story of the Cat and the Black-Bird; the Squirrel's Nest; the Equestrian Friends; and points out the Beneficent Care of Providence in implanting in the Breasts of each of his Creatures the Instinct which is necessary for its Security and Protection.

"Good evening, Uncle Thomas? We were so delighted with the adventures of Kees, that we wish to know if you have any more such amusing stories to tell us."

"Oh yes, Boys, plenty such, but it is now time to bring these Stories about Instinct to a close. I am therefore going to conclude by narrating one or two stories about the affections of animals. I wish to impress your minds with feelings of kindness towards them, and I think that the best way to do so is to exhibit them to you in their gentleness and love; to show you that they too partake of the kindlier emotions by which the heart of man is moved, and that the feelings of maternal affection, and of friendship, and of fidelity, are as much the prerogatives of the lower animals as they are of man himself. Perhaps one of the most amiable lights in which the affections of animals are exhibited is their love and attachment to their offspring. You have all seen how regardless of danger a domestic hen, one of the most timid and defenceless of animals, becomes when she has charge of a brood of chickens. At other times she is alarmed by the slightest noise—the sudden rustle of a leaf makes her shrink with fear and apprehension. Yet, no sooner do her little helpless offspring escape from the shell, than she becomes armed with a determination of which even birds of prey stand in awe."

"Oh yes, Uncle Thomas, I have often seen a hen attack a large dog and drive it away from her chickens.'

"It marks the wisdom of the omnipotent and allwise Creator, Boys, that he has implanted in the hearts of each of his creatures the particular instincts which were necessary for their safety and protection. Thus, in the case I have just spoken of, the instinctive courage with which the mother is endowed, you will find to be the best security which could have been devised. In some other birds this instinct exhibits itself in a different way. If you happen to approach the nest of the lapwing, for instance, the old birds try every means to attract your attention, and lure you away from the sacred spot. They will fly close by you, and in an irregular manner, as if wounded; but no sooner do they find that their stratagem has been successful, and that you have passed the nest unobserved, than they at once take a longer flight, and soon leave you behind."

"How very singular, Uncle Thomas! Does the lapwing defend its young with as much courage as the hen?"

"I am not aware that it does, Frank, though I think it is not at all unlikely. As its instinct teaches it to finesse in the way which I have told you, however, I should not expect to find that it does so with equal spirit. Even the pigeon, the very emblem of gentleness and love, boldly pecks at the rude hand which is extended towards its young, during the earlier stages of their existence. If you come by chance on the brood of a partridge, the mother flutters along, as if she were so much wounded that it was impossible to escape, and the young ones squat themselves close by the earth. When by her cunning wiles she has led you to a little distance, and you discover that her illness was feigned, you return to the spot to seek for the young, and you find that they too are gone: no sooner is your back turned than they run and hide themselves in some more secret place, where they remain till the well-known call of the mother again collects them under her wing.

"I lately heard a most interesting story of the

THE CAT AND THE BLACKBIRD.

boldness of a pair of blackbirds in defence of their young. A cat was one day observed mounted on the top of a railing, endeavouring to get at a nest which was near it, containing a brood of young birds. On the cat's approach the mother left the nest, and flew to meet it in a state of great alarm, placing herself almost within its reach, and uttering the most piteous screams of wildness and despair. Alarmed by his partner's screams, the male bird soon discovered the cause of her distress, and in a state of equal trepidation flew to the place, uttering loud screams and outcries, sometimes settling on the fence just before the cat, which was unable to make a spring in consequence of the narrowness of its footing. After a little time, seeing that their distress made no impression on their assailant, the male bird flew at the cat, settled on its back, and pecked at its head with so much violence that it fell to the ground, followed by the blackbird, which at length succeeded in driving it away. Foiled in this attempt, the cat a short time after again returned to the charge, and was a second time vanquished, which so intimidated her that she relinquished all attempts to get at the young birds. For several days, whenever she made her appearance in the garden, she was set upon by the blackbirds, and at length became so much afraid of them, that she scampered to a place of security whenever she saw them approach."

"That was very bold indeed, Uncle Thomas. Birds seem to be all very much attached to their young."

"Very much so, Harry; but perhaps not more so than many quadrupeds. Here is a story of the squirrel's affection, which, though it does not exhibit an instance of active defence against its enemies, affords one of endurance equally admirable.

"In cutting down some trees on the estate recently purchased by the crown at Petersham, for the purpose of being annexed to Richmond park, the axe was applied to the root of a tall tree, on the top of which was a squirrel's nest. A rope was fastened to the tree for the purpose of pulling it down more expeditiously; the workmen cut at the roots; the rope was pulled; the tree swayed backwards and forwards, and at length fell. During all these operations a female squirrel never attempted to desert her new-born young, but remained with them in the nest. When the tree fell down, she was thrown out and secured unhurt, and was put into a cage with her young ones. She suckled them for a short time, but refused to eat. Her maternal affection, however, remained till the last moment of her life, and she died in the act of affording all the nourishment in her power to her offspring.

" We are too apt, Boys, to overlook the admirable lessons which such stories as these inculcate. They should teach us kindness to each other—kindness, indeed, not only to those of our own species, but kindness to all created creatures. If the lower animals love each other so warmly and affectionately, how much more ought man, to whom the Creator has been so beneficent, to love his fellow creatures. But though the attachment of animals to their offspring is an admirable mode of its developement, it is far from being the only one. After all the Stories about Dogs—their love of their master—their fidelity—their sagacity—which I will relate to you at a future time, it is hardly necessary for me to bring forward evidence in favour of this position. Here is an instance of friendship, as it is called, between horses, which was so strong as to terminate fatally.

"During the Peninsular war, two horses, which had long been associated together, assisting to drag the same piece of artillery, and standing together the shock of many battles, became so much attached to each other as to be inseparable companions. At length one of them was killed. After the battle in which this took place, the other was picquetted as usual, and his food brought to him. He refused, however, to eat, and was constantly turning round his head to look for his companion, sometimes neighing as if to call her. All the attention which was bestowed upon him was of no avail; though surrounded by horses he took no notice of them, but incessantly bewailed his absent friend. He died shortly after, having refused to taste food from the time his former companion was killed!

"Such is but one solitary instance. But there are many such scattered up and down in the ample records of nature, bearing silent but emphatic testimony to the kindness and beneficence of the Creator. Let them but be searched for in a proper and gentle spirit, and they are sure to be found.

> "Not a tree,
> A plant, a leaf, a blossom, but contains
> A folio volume: we may read, and read,
> And read again, but still find something new—
> Something to please, and something to instruct,
> E'en in the noisome weed."

THE END.

www.ingramcontent.com/pod-product-compliance
Lightning Source LLC
Chambersburg PA
CBHW031819220426
43662CB00007B/709